Mechanical Analysis of PEM Fuel Cell Stack Design

Mechanical Analysis of PEM Fuel Cell Stack Design

Von der Fakultät für Ingenieurwissenschaften, Abteilung Maschinenbau der
Universität Duisburg-Essen
zur Erlangung des akademischen Grades

DOKTOR-INGENIEUR

genehmigte Dissertation

von

Ahmet Evren Fırat

aus
Tunceli, Türkei

Referent: Prof. Dr. rer. nat Angelika Heinzel
Korreferent: Prof. Dr.-Ing. Wojciech Kowalczyk

Tag der mündlichen Prüfung: 29.04.2016

Bibliografische Information der Deutschen Nationalbibliothek

Die Deutsche Nationalbibliothek verzeichnet diese Publikation in der Deutschen Nationalbibliografie; detaillierte bibliografische Daten sind im Internet über http://dnb.d-nb.de abrufbar.

1. Aufl. - Göttingen: Cuvillier, 2016

Zugl.: Duisburg-Essen, Univ., Diss., 2016

ISBN 978-3-7369-9257-3

eISBN 978-3-7369-8257-4

Acknowledgements

Foremost, I would like to express my sincere gratitude to my advisor Prof. Dr. rer. nat. Angelika Heinzel for her continuous support of my research and Ph.D study.

Beside my advisor, I would like to thank my second reviewer Prof. Dr.-Ing. Wojciech Kowalczyk and the rest of my thesis committee:

Prof. Dr.-Ing. Alfons Fischer

Prof. Dr. Andreas Kempf

Prof. Dr.-Ing. Stefan Panglisch

My deepest thanks also go to Dr. Peter Beckhaus for offering and leading me working on diverse exciting projects and supporting my research efforts.

I thank Michael Schoemaker for his contribution to this thesis. I am also grateful to all of my friends and colleagues at ZBT GmbH in Duisburg and University of Duisburg-Essen.

These acknowledgements wouldn't be complete without thanking my parents for their constant encouragement and support.

Table of Contents

Nomenclature

ZBT	The Fuel Cell Research Center		D	Diameter
GDL	Gas diffusion layer		T	Temperature
ΔH_H^0	Higher heating value		T_{Ref}	Reference ambient temperature
ΔG_H^0	Higher Gibbs free energy		ϕ	Potential
CHP	Combined heat power		ω	Mass fraction
mb/d	Millions of barrels per a day		R_s	Thermal contact resistance
MW	Megawatt		P_r	Prandtl`s number
PEM	Polymer electrolyte membrane		T_u	Upside contact temperature
ppm	Parts per million		T_d	Downside contact temperature
N_A	Avogadro's number		ε	Elastic strain
R_g	Gas constant		M	Molar mass
F	Faraday constant		J	Current density
E	Electrical potential		λ	Stoichiometry
R	Resistance		L_{act}	Active layer`s thickness
MEA	Membrane electrode assembly		R_{agg}	Agglomerate radius
PTFE	Polytetrafluoroethylene		D_{agg}	Gas diffusivity
POM	Polyoxymethylene		K	Henry`s constant
FEM	Finite element method		Na	Avogadro's number
CFD	Computational fluid dynamics		i_0	Exchange current density
CAD	Computer aided design		S	Specific surface area
USB	Universal serial bus		ε_{Mac}	Macroscopic porosity
x_i	Mole fraction		$n_{a,c}$	Charge Transfer number
k	Thermal conductivity		λ_{H_2O}	Waters drang coefficient
L	Characteristic length		c_{ref}	Reference concentration
Re	Reynold`s number		γ	Permeability
μ	Dynamic viscosity		φ	Electrical conductivity
Ra	Rayleigh number		θ	Compressibility factor
C_p	Heat capacity		ρ_S	Contact resistance
ρ	Fluid density		κ	Ion conductivity
p	Pressure		ϕ_{RH}	Relative humidity
BPP	Bipolar plate		T_{inf}	Ambiente temperature
I	Current		α	Thermal expansion coefficient
u	Velocity		V	Voltage

Abstract

The aim of this work is to analyze the PEM fuel cell stack design from a mechanical point of view with the help of computational methods and to develop a design procedure for fuel cell stacks. This systematic approach results in potential optimization for fuel cell stacks.

Mechanical compression of a fuel cell is required to prevent leakage and to ensure a proper electrical component contact, both of which play an important role in fuel cell performance. Fuel cells have to be held mechanically together in order to get them serially connected. Combining of fuel cells in series yields higher voltage values to deliver the desired amount of energy. The size and the number of combined single fuel cells are related to the application and requirements. The assembly of a number of single fuel cells, which is called a fuel cell stack, has to assure the identical properties for each included cell.

Compressing of fuel cells with the help of endplates and clamping them with tie rods is a conventional method used for fuel cell stack design. This method is commonly used due to the low cost and ease of fuel cell stack assembly.

Within this work a fuel cell stack of the fuel cell research center ZBT is analyzed, which is compressed with the help of endplates and tie rods. This exemplary stack has an active area of 50 cm^2 and 5 cells. The standard fuel cell stack of ZBT is focused on in this work due to the experience and wide range of examination.

Fuel cells convert chemical energy to electrical and thermal energy. The exerted thermal energy results in a temperature profile in the fuel cell stack, which affects the fuel cell mechanics via thermal expansion. The accumulating thermal expansion of the stack components results in a change in the displacement profile of the fuel cell stack. This is analyzed for ZBT fuel cell stack with the help of an appropriately prepared experimental set-up. The effect of the cell operating conditions on the mechanical properties of the fuel cell stack is also investigated with the help of measured data.

The analysis of fuel cells with the help of simulation techniques opens a gateway to product design before prototyping. It is possible to virtually investigate the behavior and properties of fuel cell systems with the help of simulations. Computational methods also enable to reach immeasurable data, which does not only facilitate the analysis, but also extends the range of design parameters.

The 3D models representing the exemplary fuel cell stack of ZBT is precisely established in accordance with the design properties. The analysis of the fuel cell stack design is carried out with the help of two computational models supplementary to the experiments. The thermomechanical characteristics of the fuel cell stack design is investigated in stack size. The effect of the compression pressure on the cell performance is examined on a cross-section model by implementation of the electrochemistry and fluid dynamics.

The thermomechanical model of the stack on a large scale contributes to the understanding of the fuel cell stack design from a mechanical point of view and leads to the dimensioning of stack components properly. Thermal expansion is to be implemented into the structural mechanics for a proper mechanical analysis as realized by the performed measurements. The computation of thermal expansion for each component is taken into account by integrating cell operating temperature profiles into the structural analysis.

The cross-section model is derived precisely from the fuel cell supplemental to the stack model. The effect of the compression pressure on the fuel cell dynamics is analyzed with the help of this cross-section model including electrochemistry. The polarization curve as a result is examined in detail. The sensitivity analyses are performed to understand the effect of selected parameters on the fuel cell performance and the dynamics of simulations.

As a result the mechanical pressure distribution profile and its effect on the fuel cell performance are analyzed. Performed measurements are utilized for the verification of the computational models. The value of compression pressure in the fuel cell must be small enough to assist the gas diffusion in the GDL. Beside this, it must also be large enough to fulfill the requirements for a proper contact resistance and tightness.

The explicit mechanical analysis of the standard fuel cell stack of ZBT with the help of experimental and computational methods leads to an optimization procedure. The design procedure developed in this study can be used for different sizes of fuel cell stack designs and similar applications like electrolysis or flow batteries.

Zusammenfassung

Im Rahmen dieser Arbeit werden die mechanischen Analysen eines PEM-Brennstoffzellenstacks sowie die Entwicklung einer Methodik für die Auslegung eines Brennstoffzellenstacks mit Hilfe von den rechnergestützten Simulationen durchgeführt. Diese systematische Methodik ermöglicht die Optimierung des Brennstoffzellenstackdesigns.

Die mechanische Verpressung der Einzelzellen ist erforderlich für die Gasdichtigkeit und die elektrische Kontaktierung der Komponenten. Dies spielt eine wesentliche Rolle für die Leistung der Brennstoffzellen. Brennstoffzellen müssen mechanisch verspannt werden, um deren serielle Schaltung zu erstellen, die zu den höheren Spannungswerten und demensprechend zur benötigten Energiemenge führt. Die Größe und Anzahl der gestapelten Brennstoffzellen richten sich dabei nach der Zielanwendung und den Systemanforderungen. Ein Brennstoffzellenstack muss diese Anforderungen für jede einzelne Zelle sicherstellen.

Die Verspannung der Brennstoffzellen unter Verwendung von Endplatten und Zugstangen ist ein konventioneller Aufbau für die Brennstoffzellenstacks. Dieser Aufbau wird auf Grund seiner leichten und kostengünstigen Ausführung häufig in Brennstoffzellenstacks angewendet.

Im Rahmen dieser Arbeit wurde ein aus 5 einzelnen Zellen mit einer Fläche von jeweils 50cm2 bestehender ZBT-Brennstoffzellenstack, welcher mittels Endplatten und Zugstangen verspannt wurde, im Detail untersucht. Der Standard-Brennstoffzellenstacks des ZBT wurde ausgewählt, weil zu diesem Stack bereits umfangreiche Untersuchungen am ZBT durchgeführt wurden.

In Brennstoffzellen wird chemische Energie in elektrische und thermische Energie umgewandelt. Die entstandene Wärme führt zu einer Temperaturverteilung innerhalb des Brennstoffzellenstacks. Diese beeinflusst die mechanischen Eigenschaften des Brennstoffzellenstacks durch die thermische Ausdehnung. Die gesamte thermische Ausdehnung aller Komponenten innerhalb des Brennstoffzellenstacks ändert die Auslenkung des Brennstoffzellenstacks im Betrieb. Dies wurde innerhalb dieser Arbeit für einen Standard-Brennstoffzellenstack des ZBT mit der Hilfe eines entsprechend eingerichteten Versuchsaufbaues untersucht. Der Einfluss von Betriebsparametern auf die mechanischen Eigenschaften des Brennstoffzellenstacks wurde mit der Hilfe der aufgenommenen Messdaten analysiert.

Die Analyse der Brennstoffzellen mit Hilfe der Berechnungsmethoden ermöglicht Zeit- und Kostenreduktionen in der Entwicklungsphase. Durch Simulationen können Brennstoffzelleneigenschaften virtuell untersucht und optimiert werden. Berechnungsmethoden erlauben die Evaluierung der am Prüfstand messtechnisch nicht erfassbaren Parameter. Im Rahmen dieser Arbeit wurden zwei 3-dimensionale Modelle des exemplarisch verwendeten ZBT-Brennstoffzellenstacks entsprechend der Designeigenschaften entwickelt. Die Analyse des Brennstoffzellenstacks wurde mit der Hilfe von zwei Berechnungsmodellen ergänzend zu den Messdaten durchgeführt. Die thermomechanischen Eigenschaften eines aus 5 einzelnen Zellen bestehenden Brennstoffzellenstacks wurden mit dem ersten Modell charakterisiert. Der Einfluss der Anpresskraft auf die Leistung der Brennstoffzelle wurde mit Hilfe eines zweiten Schnitt-Modells unter Berücksichtigung der Elektrochemie und Strömungsmechanik untersucht.

Im Hinblick auf die mechanischen Eigenschaften können die Designparameter eines Brennstoffzellenstacks mit Hilfe des Stackmodells im Detail analysiert werden. Dies führt zur Analyse der Dimensionierung von Stackkomponenten. Durch die Evaluierung der Messdaten soll die thermische Ausdehnung für eine explizite mechanische Analyse des Brennstoffzellenstacks bei der mathematischen Modellierung betrachtet werden. Die Berechnung der thermischen Ausdehnung wurde mit der Implementierung der simulierten Temperaturverteilung aller Komponenten in strukturmechanischen Modellierungsansatz durchgeführt.

Ergänzend zu dem Stack-Modell wurde das Schnitt-Modell mit Hilfe der Geometrie der Brennstoffzellenkomponenten abgeleitet. Der Einfluss der Anpresskraft auf die elektrochemischen Eigenschaften der Brennstoffzellen konnte mit Hilfe dieses Schnitt-Modells untersucht werden. Die von den Simulationen abgeleitete UI-Kennlinie wurde im Detail analysiert. Die Einflüsse der ausgewählten Betriebsparameter auf die Leistung der Brennstoffzellen konnten mit der Hilfe einer Sensitivitätsanalyse untersucht werden.

Im Rahmen dieser Arbeit wurden die Kraftverteilung innerhalb des Brennstoffzellenstacks und deren Einfluss auf die Zellleistung im Detail untersucht. Die aufgenommenen Messdaten wurden für die Verifizierung der Berechnungsmethoden verwendet. Für die Anpresskraft existiert dabei ein Optimum: Für die elektrische Kontaktierung und Gasdichtigkeit ist einerseits eine maximale Anpresskraft erforderlich, anderseits wird die maximale Anpresskraft durch eine reduzierte Gasdiffusion innerhalb der Brennstoffzellenkomponenten limitiert.

Durch die Analyse eines ZBT-Brennstoffzellenstacks unter Verwendung von Simulationen und Experimenten an realen Stacks konnte eine Methodik für die optimale Auslegung eines Stacks erstellt werden. Diese Methodik kann sowohl für die Optimierung von Brennstoffzellenstacks unterschiedlicher Größen, als auch für die Auslegung von ähnlich aufgebauten Elektrolyseuren und Flow-Batterien verwendet werden.

1 Motivation

The use of fossil fuels to cover the energy demand in the future carries key problems to overcome. The environmental and economic concerns regarding the climate change and the decrease in the oil reserves lead the research and industry to the search for alternatives [1]. The increase in the energy demand due to the economic and demographic growth in the world enhances the necessity of these research efforts. The dependency of several developed countries on the fossil fuel economy and political instability for energy supply enforce the search for the alternatives additionally.

The global energy economy is not going to depend on only one single energy source in order to get a secure energy supply. The product range for the energy supply differs from nuclear power to geothermal energy. But renewable energy sources fulfill the requirements as a long term solution. They are getting incrementally more importance for the energy markets due to the depletion of fossil fuels and environmental concerns. The predicted growth in supply of renewable energy sources reaches an average value of 8.1% p.a. until 2035 [2].

The fluctuations in renewable energy production such as solar, wind and hydropower prevent sustainable energy sources to provide a stable solution to the energy problems [3-7]. Storing the energy produced by renewable sources is a key phenomenon in order to remove production instabilities and is focused by research and industry in last decade in order to achieve the global energy goals. The production and storage of renewable energy are significant issues in order to increase the use of renewable energy applications.

Production of hydrogen is considered as an alternative way for storing the electrical energy produced by renewable sources. Production of hydrogen with electrolysis and storing the produced hydrogen in special tanks or underground are actual research efforts and open the gateway to the hydrogen economy. The hydrogen economy is also feasible via producing the hydrogen from fossil fuels by reforming processes but this doesn't offer a long term solution.

Fuel cell systems represent a potentially viable option with high level of efficiency for re-electrification of hydrogen [8]. Using fuel cells to utilize hydrogen has high efficiency values depending on the type of the fuel cell [9]. The PEM type fuel cells promise several advantages for fuel cell applications due to the low temperature requirements. Hydrogen can also be converted to electrical energy by using a combustion process. However this reduces the overall efficiency [10].

Several PEM fuel cell systems have been demonstrated in the last decade. Besides the hydrogen supply and the costs for the fuel cell systems, the durability of fuel cells plays an important role for the commercial launch. The mechanical stability of fuel cells leads to a higher level of lifespan due to the durability of single fuel cell components like membrane and gas diffusion layers [11, 12]. It is also required to get an optimal mechanical pressure on the fuel cell components in order to get an optimal efficiency from the fuel cell stack.

The overall aim of this work is to enable the explicit analysis of fuel cell stack designs from a mechanical point of view in order to increase the mechanical durability and reach the optimal performance. Understanding the mechanical characteristics of the fuel cell stacks leads to the dimensioning of the stack components and contributes to the optimization procedure during the prototyping process, which assists in the cost reduction of fuel cell stacks. This also enables to make safe and compact fuel cell stack designs depending on the application, which plays an important role for implementation of fuel cells for stationary and transportation applications. The developed design procedure in this study and the analysis for a specific fuel cell stack can also be implemented both for other fuel cell stacks and similar applications.

2 Introduction

A fuel cell is a device, which converts chemical energy directly to the electrical energy in a single electrochemical step with the help of catalyst. In principle a fuel cell operates similar to a battery. A fuel cell does not require recharging in comparison to a battery. It produces energy in the form of electricity and heat as long as fuel is supplied.

Different from Carnot heat engines, the efficiency of the fuel cells reaches to the theoretical values of $\eta = 83.3\%$ owing to the elimination of the mechanical energy conversion process [9]. There are no moving parts in the fuel cells involved in the production process of electricity which contributes to the higher level of efficiency. A single fuel cell consists of two electrodes sandwiched around an electrolyte layer and produce electrical current principally by consuming hydrogen and oxygen.

2.1 *Fuel Cell Applications*

Fuel cells are used for several stationary, transportation and portable applications. The power range generated by fuel cells can differ from milliwatts to hundreds of kilowatts. Additionally there are several types of fuel cells with different operating temperature levels and properties (see section 2.2.1) leading to an extended area of potential applications. Principally fuel cells can be used for any power generation application. In this section some selected applications are categorized and given in subsections to provide an overview of the wide range of fuel cell applications and required conditions in order to understand the mechanical requirements of fuel cells.

2.1.1 Stationary Applications

The stationary fuel cell applications denote power generation units for homes, buildings and auxiliary power units etc. The power range of stationary fuel cell applications can be between 0-500kW [19]. Some types of fuel cells are commonly used for stationary applications due to their specific properties as handled in section 2.2.1.

The low dynamic response requirement is the basic characteristic property of stationary applications. Another common property of stationary applications is the usage of natural gas due to the utilization of existing natural gas infrastructure. Two selected examples of different sized stationary applications are given in Figure 2.1 and Figure 2.2. The size, operating conditions and application areas of stationary fuel cells are various, which contributes to diverse mechanical requirements.

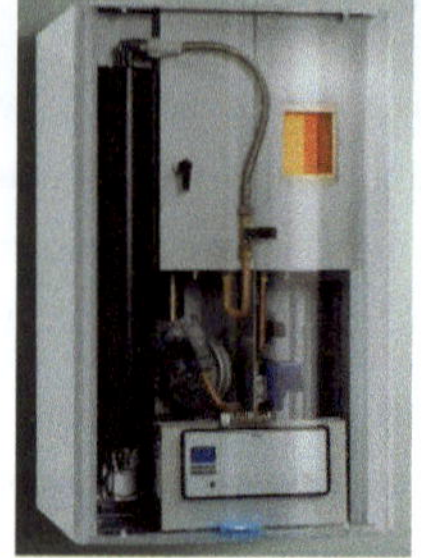

Figure 2.1 CalTech, Bloom Energy (Image Courtesy Bloom Energy)

Figure 2.2 Vaillant CHP System (Image Courtesy Vaillant GmbH & Co.KG)

2.1.2 Portable and Transportation Applications

Batteries don't promise sufficient fast recharging time and energy density for long-term operations both of which are key properties for portable and transportation applications. Fuel cells assure a viable option for battery applications in small power range between 0-100W [19]. The implementation of fuel cells in transportation is one of the foremost fuel cell applications. Dynamic response requirements are the main characteristic property of portable and transportation applications. Volume and weight play also an important role as design parameters. Two selected examples from portable and transportation applications are given in Figure 2.3 and Figure 2.4. The product range differs from small sized fuel cells for the recharging of mobile phones to bigger sized fuel cells for power train in automotive applications. The mechanical requirements of these fuel cell applications depend on the type of the fuel cell, size of the fuel cell stack and application based specifications, which can be varied distinctively.

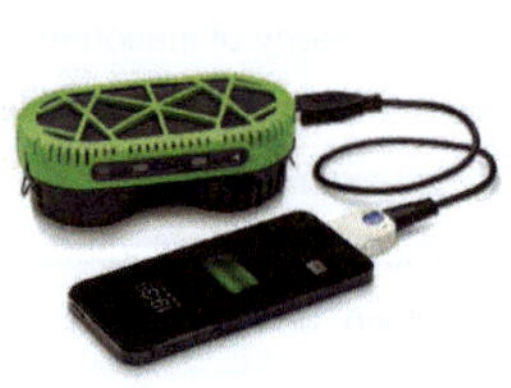

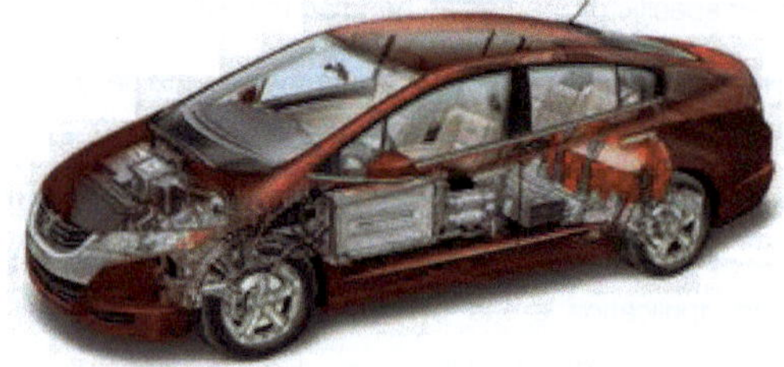

Figure 2.3 Portable fuel cell application

(Image Courtesy PowerTrekk)

Figure 2.4 Fuel cell car (Image Courtesy Honda Motors)

2.2 Theoretical Background

Basic theory and structure of fuel cells regarding the issues of this work are handled in this section. For further details it is referred to literature [9, 10, 18-21].

2.2.1 Type of Fuel Cells

Fuel cells can be classified as given in Table 2.1 based on working principle and operating temperature.

Table 2.1 Type of Fuel Cells and Their Properties

Type of Fuel Cell	Electrolyte	Operating temperature	Power level and electrical efficiency	Fuel Transferring ion	Application areas
AFC Alkaline fuel cell	Potassium hydroxide (KOH)	50-200°C	10-100kW 60-70%	Hydrogen OH⁻	Space, power generation, military
PEM Polymer electrolyte membrane fuel cell	Sulphonic acid incorporated into a solid membrane	50-90°C	0.01-1000kW 50-68%	Hydrogen H^+	Transport, power supplies CHP, space, military
DMFC Direct methanol fuel cell	Sulphonic acid incorporated into a solid membrane, or Sulphuric acid solution (Nafion, Dow)	50-110°C	0.001-100kW 20-30%	Methanol H^+	Portable electronic systems, mobile consumer electronics
PAFC Phosphoric acid fuel cell	Phosphoric acid (H_3PO_4)	190-210°C	100-5000kW 55%	Hydrogen H^+	CHP, power generation
MCFC Molten carbonate fuel cell	Molten lithium carbonate (Li_2CO_3, K_2CO_3)	630-650°C	1-100MW 65%	Hydrogen CO_3^{2-}	Large stationary power
SOFC Solid oxide fuel cell	Ceramic, solidoxide, zironia (ZrO_2/YO_3)	800-1000°C	0.1-100MW 60-65%	Hydrogen O^{2-}	CHP power generation, transport

There are various fuel cell types mainly categorized with respect to electrolyte, which bases on different transferring ion types. Each of them has distinct advantages and disadvantages upon other ones such as starting time, operating temperature, efficiency and sensitivity to the fuel purity. This leads to different application areas as seen in Table 2.1.

PEM fuel cells have low operating temperature, which makes this type of fuel cells appropriate for a wide range of applications as it can be seen in Table 2.1. Therefore, PEM fuel cells have been widely focused on by researchers because of its high power density and short starting time due to the low temperature requirements. These make PEM fuel cells remarkable also for transportation applications.

2.2.2 Thermodynamics of PEM Fuel Cells

Basically fuel cells are working in a reverse principle of electrolysis. They consume oxygen and hydrogen and produce water and energy in terms of heat and electricity. The basic reactions of a PEM fuel cell can be seen in Eq. 2.1-3. The reactions in Eq. 2.1 and Eq. 2.2 take place at the anode (-) side and cathode (+) side respectively [9, 10].

$$H_2 \Rightarrow 2H^+ + 2e^-$$

Eq. 2.1

$$\frac{1}{2}O_2 + 2H^+ + 2e^- \Rightarrow H_2O$$

Eq. 2.2

The overall reaction at cathode side is the sum of both reactions (see Eq. 2.1 and Eq. 2.2) and can be seen in Eq. 2.3.

$$H_2 + \frac{1}{2}O_2 \Rightarrow H_2O + 285.8kj/mol$$

Eq. 2.3

The overall reaction is an exothermic reaction and releases an amount of energy as it can be seen in Eq. 2.3. The higher heating value of hydrogen $\Delta H_H^o = 285.8kJ/mol$ is taken into account to calculate the maximum amount of available energy by using liquid water as product [9, 10]. The total amount of available energy cannot be converted to the electrical energy by fuel cell due to the entropies occurring during the reaction. The maximum amount of electrical energy generated in a fuel cell corresponds to the Gibbs free energy ΔG_H^o.

Therefore the overall theoretical fuel cell efficiency can be calculated as in Eq. 2.4 with the help of Gibbs free energy ΔG_H^o.

$$\eta = \frac{\Delta G_H^O}{\Delta H_H^O} = \frac{237.3kJ/mol}{285.8kJ/mol} \times 100 = 83\%$$

Eq. 2.4

The theoretical potential is calculated as in Eq. 2.5 assuming that all of the Gibbs free energy is converted into electrical energy. n is the number of the participating electrons in the reaction (for $H_2 = 2$). F represents the Faraday's constant ($F = 96,485 \frac{C}{mol}$), which is a product of the number of molecules per mole (Avogadro's number, $N_A = 6.022 \times 10^{23}$) and charge of 1 electron ($e = 1.602 \times 10^{-19}C$).

$$E = -\frac{\Delta G_H^O}{n \times F} = 1.23V$$

Eq. 2.5

2.2.3 Operating Principles of PEM Fuel Cells

A single PEM fuel cell has the structure as illustrated in Figure 2.5. The hydrogen consumption side of a fuel cell is negatively charged and called anode. Contrary to anode side, the oxygen consumption side is called cathode and positively charged. A fuel cell consists of numerous components as shown in Figure 2.5. Each of them has significant properties, those affect directly the efficiency of the fuel cell.

The central component of the PEM fuel cell is the membrane, which acts as electrolyte in the fuel cell. The membrane is impermeable to gases and has the capability of transmitting protons. The protons are transferred from the anode to cathode side through the membrane and the electrons are delivered to the cathode side through the external circuit by which it produces electricity as illustrated in Figure 2.5. At the cathode side of fuel cell the protons reunite with oxygen ions to water molecules. The solid state of the membrane at the fuel cell operating conditions assures several advantages e.g. handling and assembling.

The reactions defined in section 2.2.2 take place at the catalyst layers which are located on the surfaces of membrane and consist of platinum particles bonded with carbon nanoparticles.

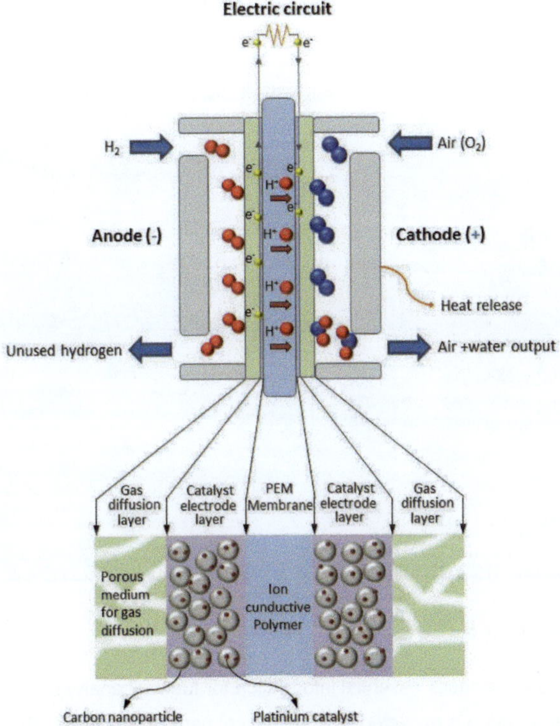

Figure 2.5 The structure and molecular transportation of a PEM fuel cell

The membrane is sandwiched between two porous layers as seen in Figure 2.5, those are electrically and thermally conductive. The porous layers are commonly called GDL (Gas Diffusion Layer) and made from carbon cloth or carbon fiber paper. They are acting as electrodes in fuel cells and have the tasks of gas distribution and transferring the electricity produced in the fuel cell. The porous structure of GDL ensures the uniform distribution of the gases on the entire catalyst surface.

2.2.4 Electrochemistry of Fuel Cells

The basic electrochemical reactions and thermodynamics of a PEM fuel cell are briefly explained before. The polarization curve and electrochemical losses are described in this section to understand the electrochemical characteristics of fuel cells. Further information can be found in literature [9-13, 18, 19].

The polarization curve represents the cell current-voltage relationship and is the most important characteristic property of a fuel cell to evaluate the cell performance. A typical polarization curve is given in Figure 2.6. Mostly the current is also scaled on the membrane active area to get the current density. The current density is a standard comparable quantity for cell evaluation.

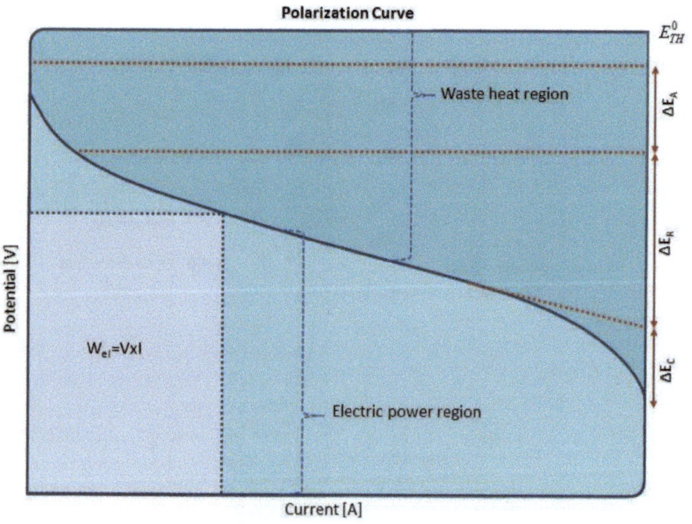

Figure 2.6 Polarization Curve of a fuel cell, schematic diagram

The polarization curve releases essential information for fuel cell analysis. As seen in Figure 2.6, a drop in voltage value is expected as a function of generated current. This occurs due to the irreversible internal losses.

There are several important elements which induce voltage losses during operation of a fuel cell. The three basic ones are ΔE_A, ΔE_R, ΔE_C and illustrated in three different regions in the Figure 2.6. They are also listed in Table 2.2 and analyzed further in the following sections.

Table 2.2 Basic types of voltage losses

ΔE_A	Activation losses
ΔE_R	Resistive losses
ΔE_C	Concentration losses

Understanding the potential losses enables the analysis of the polarization curve structure and the electrochemical behavior of the fuel cell. It is required to conceive the basics of the simulations performed in section 4.3.

The cell potential can be defined as in Eq. 2.6. E_{TH} is the theoretical potential and given as 1.23 [V] in Eq. 2.5.

$$E_{cell} = E_{TH} - \Delta E_A - \Delta E_R - \Delta E_C \qquad \text{Eq. 2.6}$$

The area under the polarization curve represents the electric power region. The rest of the available energy is released as heat as depicted in Figure 2.6 and can be calculated by cell voltage value as dealt in section 2.2.5.1.

2.2.4.1 Activation Loss

Under low current operating conditions the activation losses dominate the cell potential as shown in Figure 2.6. The activation loss represents the voltage drop required to initiate the reactions. The activation losses occur at both anode and cathode side of the fuel cell. The cathodic overpotential is significantly higher than the anodic overpotential. Neglecting the anodic overpotential, the activation loss can be defined as in Eq. 2.7 [9, 10].

$$\Delta E_A = \frac{RT}{\alpha F} ln \left(\frac{i}{i_0} \right) \qquad \text{Eq. 2.7}$$

R and T represent the gas constant and temperature respectively. α and F are transfer coefficient and Faraday's constant. i and i_0 represent the current density and exchange current density respectively.

2.2.4.2 Resistive Loss

The voltage value drops in a linear behavior at moderate current density region as shown in Figure 2.6. The main contribution is the ohmic resistance of the electrolyte caused by limited proton conductivity. The electrical resistance of the cell components induce an additional voltage drop. The contact between electrical conductive fuel cell components leads to the contact resistance. These result in an overall potential loss which can be calculated using Ohm's law as defined in Eq. 2.8.

$$V = IR_0$$

Eq. 2.8

The values of electrical and ionic resistance of fuel cell components can be found in section 4.3.2.4.

2.2.4.3 Concentration Loss

The concentration loss denotes the reduction in the reactant concentration on the membrane surfaces due to rapid consumption. The reactant supply on the electrolyte surfaces reaches to its limits to cover the reaction rates at higher level of current densities. This results in a potential drop. Concentration losses dominate the voltage drop at higher current densities and form the tail of the polarization curve as depicted in Figure 2.6.

2.2.5 Operating Conditions of PEM Fuel Cells

Operating conditions of a PEM fuel cell can be specified due to the type of application. Special components and control systems are required to preset the conditions for gas flow, humidifying, cooling etc., mainly for achieving the required power but also the dynamic response characteristics of a fuel cell. The basic operation conditions of a fuel cell are expanded in this section with the interest in understanding of the fuel cell dynamics and simulations made in chapter 4.

2.2.5.1 Operating Temperature

As mentioned before fuel cells release energy in the form of heat and electricity. The amount of available energy which cannot be converted to electrical power is released as heat, which is to be removed from the system to get a stabilized temperature profile in the fuel cell. The heat removal is performed by using a cooling system which depends on the system requirements. Cooling systems of fuel cells are discussed in detail in chapter 2.3.

The optimal operating temperature of a PEM fuel cell can be between 50°-90°C as it can be extracted from Table 2.1. Higher operating temperatures in fuel cells result mainly in higher fuel cell power and efficiency [10, 14, 15]. Nevertheless the highest applicable temperature cannot be inferred due to degradation of the fuel cell [16, 17]. For each fuel cell design and application there is an optimal temperature range. The operating temperature of the standard 50 cm^2 fuel cell design of ZBT is about 70°C.

The generated heat power (Q_{gen}) from the fuel cell can be extracted from Eq. 2.9 [10, 13]. ET_H^0 is the theoretical thermoneutral potential derived from the hydrogen's higher heating value under complete condensation of product water and has a value of 1.482 [V]. V_{cell} and I

represent the measured cell voltage and the current respectively. n_{cell} is the number of cells in the fuel cell stack.

$$Q_{gen} = (E_{TH}^0 - V_{cell})I \cdot n_{cell}$$

Eq. 2.9

The temperature profile of a fuel cell stack plays an important role from the mechanical point of view due to the thermal expansion occurring in the fuel cell stack. The temperature distribution and the thermal stresses occurring in the fuel cell stack are handled precisely in section 4.2.

2.2.5.2 Relative Humidity

The polymer electrolyte membrane is proton conductive as mentioned before. The conductivity of the membrane strongly depends on humidity. Because of this special property of the membrane, the reactant gases must be humidified before entering the fuel cell to get uniform proton conductivity. The humidity of reactants can differ for anode and cathode sides distinctively depending on the current and other operating conditions such as temperature. The water is additionally produced as product at the cathode side and is to be removed from the fuel cell. Feeding the membrane with more water than required can cause undesirable flooding of the fuel cell channels or porous GDL and electrode structures with liquid water. This prevents the gases from flowing through the channels to reach the electrode surface and decreases the fuel cell efficiency. The required relative humidity for the reactants depends on the operating points and fuel cell structure. This optimal humidity value can be obtained from measurements. Further detailed information about the membrane can be found in section 2.3.2.1.

2.2.5.3 Operating Pressure

The operating pressure of a fuel cell can be varied depending on the specification of application and components. Fuel cells can also be operated at ambient atmospheric conditions. Higher pressure conditions correspond to higher values of potential, power and fuel cell efficiency. For specific applications such as automotive, it is required to get higher energy density from the fuel cell, which leads to higher pressure requirements. Operating fuel cells at high pressure requires additional fans or compressors to be used, which results in additional energy consumption. But as mentioned, it is feasible for some applications to accept a decrease in system efficiency by using additional components to pressurize the reactants. The operating pressure affects the pressure distribution in the flow field, which is required for the simulations performed in section 4.3 for the calculations of gas concentration distribution.

2.3 *Fuel Cell Stack*

In order to get higher voltage values, the single fuel cells have to be electrically connected in series. The assembly of stacked fuel cells to establish this serial electrical configuration is called fuel cell stack. There are several types of fuel cell stack designs [22-33]. Three selected fuel cell stacks can be seen in Figure 2.7-9. The requirements to be considered by performing the stack design are identical for all stack concepts. The structure and basic principles of fuel cell stacks are briefly handled in this section focusing on the standard fuel cell stack design of ZBT seen in Figure 2.9.

Figure 2.7 Fuel cell stack (Image Courtesy Intelligent energy)

Figure 2.8 Fuel cell stack (Image Courtesy Honda Motors)

Figure 2.9 Fuel cell stack of ZBT (Zentrum für BrennstoffzellenTechnik)

2.3.1 ZBT Fuel Cell Stack Design

Standard fuel cell stack design of ZBT can be seen in Figure 2.9. The ZBT fuel cell stack design and its electrical circuit configuration are depicted in Figure 2.10.

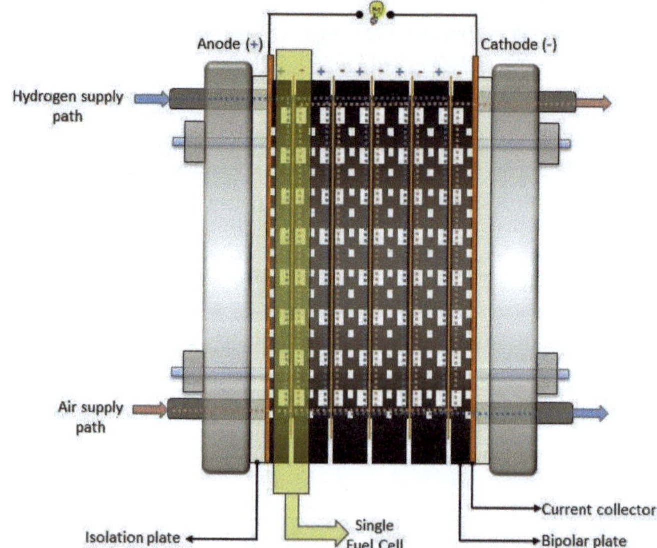

Figure 2.10 Fuel cell stack and electrical circuit components

The fuel cell stack provides the medium supply for each single cell through the flow paths illustrated with dashed lines in Figure 2.10. Fuel cell stack provides the electrical conduction and the tightness of the flow paths via mechanical compression. The required operating conditions handled in section 2.2.5 are to be assured for each single cell in the fuel cell stack.

A cooling system is to be integrated in the fuel cell stack to maintain the desired temperature level. In order to prevent drying out of the membrane and excessive thermal stresses, an adequate temperature level is required [34]. There are several existing cooling techniques from evaporative to air cooled systems [35-37]. The fuel cell stack with illustrated cooling fins in Figure 2.10 demonstrates an air cooled system. An air cooled fuel cell stack is investigated by the measurements and performed simulations in section 3 and 4.2 respectively.

The key points to be considered by fuel cell stack design can be listed as below:

- Preventing leakages
- Minimizing the electrical losses
- Thermal management
- Mechanical stability
- Uniform and sufficient medium supply
- Cost and specification of application

For further details about design of fuel cell stack it is referred to literature [9, 18, 19].

2.3.2 PEM Fuel Cell Stack Components

The components used in standard 50 cm^2 fuel cell stack of ZBT are illustrated in Figure 2.11 as exploded assembly view and categorized in Table 2.3. Each component has significant functions affecting cell performance. In the following sections the basic stack components and their characteristic properties are concisely handled.

Table 2.3 Fuel cell components

Fuel cell components		
	Medium suppliers	Bipolar plate
		Gas connector
	Mechanical clamping components	Endplate
		Tie rod
		Disc spring
		Flat washer, screw nut
		Pressure plate
	Cell components	MEA
		GDL
		Sealings
	Stack components	Current collector
		Isolation plate

The material properties of the stack components are given in detail by performing the calculations made in section 4.2.

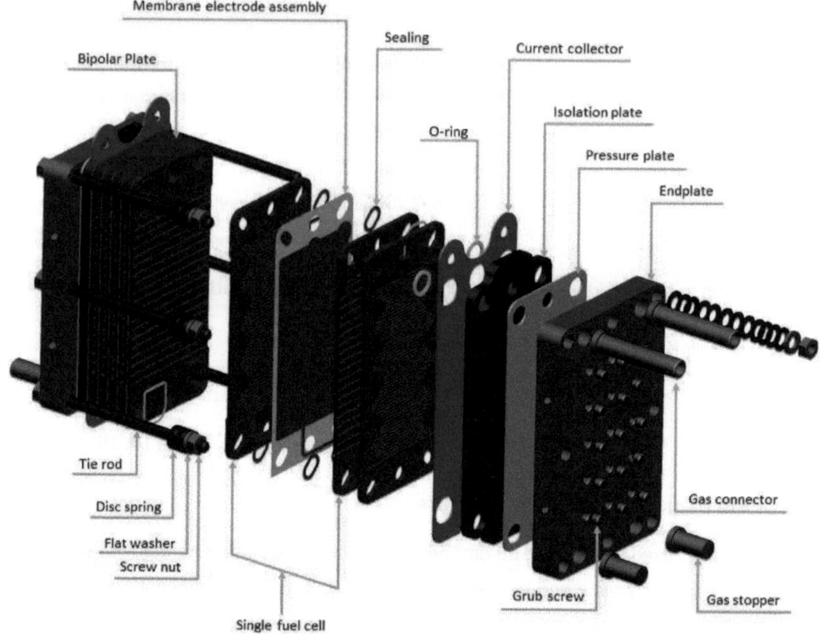

Figure 2.11 Fuel cell stack and components

2.3.2.1 Cell Components

The MEA (Membrane Electrode Assembly) consists of a membrane sandwiched between two electrode layers (see Figure 2.5). The basic characteristics and requirements for MEA are given in sections 2.2.3 and 2.2.5. The mostly used material for PEM fuel cell membrane is PTFE (polytetrafluoroethylene). The basic property of the membrane is the capability of proton conductivity and preventing the mixing of the reactants. It must also be mechanically and chemically stable for proper cell operation and lifespan, which has to be regarded due to the thin structure of the membrane.

O-Ring gaskets and dispensed sealings are used to get the tightness of the gas connectors and fuel cells respectively. Carbon paper type of GDL and silicone as sealing material are considered as standard ZBT cell components by calculations made in section 4. The detailed information about the cell components and their characteristic properties can be found in [10, 18, 19].

2.3.2.2 Medium Suppliers

The gas connectors are the components used for the supply and removal of the flow media. The contact between gas connector and media limits the material selection of the gas connectors due to the sensitivity of cell components. Metal ions originating from gas connectors can cause degradation of membrane [38]. POM (Polyoxymethylene) is used as material for the gas connectors and stoppers in standard ZBT fuel cell stack considering the membrane degradation, machinability and material properties.

The bipolar plates are multifunctional fuel cell components acting as current collector, fuel distributor and media separator. The design and material of bipolar plates determine the functionality of a fuel cell stack. The required characteristic properties for bipolar plate material are electrical and thermal conductivities, mechanical stability and corrosion durability [39]. Metals and graphite based compounds are mostly used as bipolar plate material. Metallic bipolar plates have several advantages as well as disadvantages [40, 44]. The detailed information about metallic bipolar plates can be found in [41-44]. Graphite based compounds exhibit excellent corrosion resistance but poor machinability. Injection moulding of graphite based compounds is a feasible manufacturing method for bipolar plates [45, 46]. Injection moulded graphite based bipolar plates of ZBT are focused on in this work. Further information about bipolar plates can be found in literature [10, 39, 44, 47-49]. The standard bipolar plate design of ZBT can be seen in Figure 2.12.

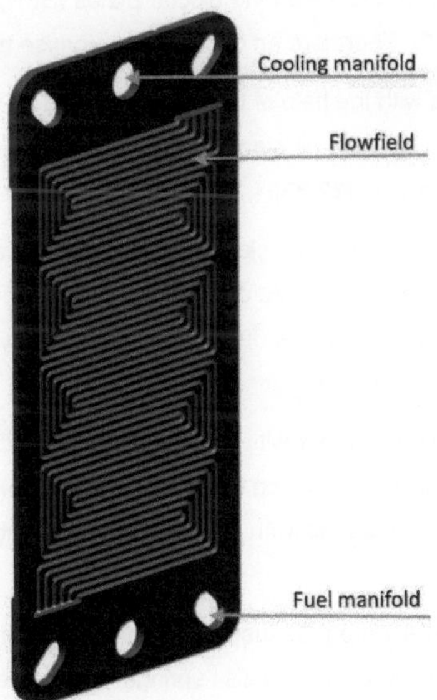

Figure 2.12 Standard bipolar plate design of ZBT

The design of bipolar plate is adapted to the injection moulding properties of graphite composite. The utilization of the whole active membrane area is also taken into consideration by the design of the flow field structure. The manifolds and serpentine flow field of the bipolar plate can be noticed in Figure 2.12. The flow channel structure and the width of the ribs on the flow field are developed due to the experimental analysis.

The dimensioning of the bipolar plates is critical for the stack properties. The thickness of the bipolar plate is a significant parameter, which is to be considered due to the injection molding properties and thermal performance. Thicker bipolar plates result in improved molding properties and mechanical stability. Nevertheless it denotes the increase of the total weight of the fuel cell stack and loss in the thermal and electrical performance of the bipolar plate. The graphite based compound material and its injection moulding properties as well as the design of the bipolar plates are handled in [50, 51].

2.3.2.3 Mechanical Clamping Components

The main aim of the mechanical clamping components is to realize the tightness of the fuel cell stack assuring the minimum contact resistance and uniform internal stress distribution.

The requirement for tightness and low contact resistance result in high compression pressure values for stack design. However over-compressing of fuel cell components such as the gas diffusion layer (GDL) disturbs their porous structures thus reduces the performance of a fuel cell [52, 53]. It can also cause cracks on the bipolar plates (BPP) due to the brittleness of the graphite based material [53]. Therefore an optimal compromise has to be achieved.

Compressing the fuel cells with the help of two endplates and clamping them with tie rods is a conventional method used in fuel cell stack design considering its simplicity and low costs compared to other stack designs [22-33].

The endplates, as the main mechanical clamping components, are responsible for the even distribution of the applied clamping forces on the fuel cells. Aluminum alloy (5083) is used as standard material for the ZBT endplates due to the machinability, low cost and relatively low weight.

The tie rods (DIN 975) acting together with flat washers (DIN 125) and screw nuts (DIN 934) are used for applying the clamping force on the endplates. The calculation of the clamping load acting on tie rods can be performed with the help of tightening torque of the screws and handled in section 4.2.2.3.

The disc springs (DIN 2093-A) are to be used in order to compensate the thermal expansion during operation of the fuel cell stack. The disc springs have a special arrangement in order to get a proper spring characteristic curve for ensuring the identical acting force under operating conditions.

The pressure plate (stainless steel 1.4301) is used for applying additional mechanical pressure directly to the cells without acting on the endplates. Additional mechanical pressure can be applied with the help of grub screws (DIN 913) embedded in endplates. The use of grub screws increases the design complexity and cost. Grub screws are commonly not required and neglected by the simulations performed in section 4.2.

The design principles of the mechanical clamping components are identical for several fuel cell stacks. The geometry and size of the clamping components are formed due to the bipolar plate design and specification of the application. The tightness of the fuel cell stack is to be considered by dimensioning of the mechanical components. An appropriate dimensioning of the mechanical components corresponds directly to the weight of the fuel cell stack, which has to be considered for a proper stack design.

2.3.2.4 Stack Components

The current collector is a copper plate used to transmit the produced current from the fuel cell stack to an electric circuit via corresponding terminals. The terminals of the current collector are under mechanical load due to the cabling. In order to prevent an undesired deformation of the current collector, an appropriate dimensioning of the current collector is performed according to the material properties of copper. Dimensioning of the current collector also influences packaging of the fuel cell stack in a system. The current collector is nickel-plated and gold-coated in order to avoid contact resistances. The gold and nickel coatings of the current collectors are neglected by the calculations made in section 4.2 regarding the irrelevance from a mechanical point of view.

The isolation plate is used to achieve thermal and electrical insulation. It prevents the thermal energy to reach the endplate and thus the terminal cells to stay cooler compared to cells in the center. The thermal energy exerted in the fuel cells is utilized together with a cooling system to get the desired temperature values for a proper cell operation. Polyoxymethylene (POM) is used as standard material for the isolation plates in ZBT fuel cell stacks regarding its functionality, costs and machinability. The isolation plate is also utilized for fixing the gas connectors and stoppers by stacking of the fuel cells.

Isolation plate is used together with the endplate for the positioning of the fuel cell components. The thin structure of the cell components makes their handling and assembling of the stack complicated. Assembling of the fuel cell stack plays an important role for commercial launch due to the manufacturing costs and has to be taken into consideration for a proper fuel cell stack design.

The design and the dimensioning of the components of the ZBT stack and their material selections have been developed due to the wide range of experimentations. A proper analysis of the standard ZBT fuel cell stack hasn't been performed from a mechanical point of view, but could lead to a better understanding and optimization of the fuel cell stack design. This also builds the basis for the analysis of any other PEM fuel cell stack design. The analysis is performed in the following sections with the help of experiments and computational methods.

3 Cell and Stack Tests

In order to study the standard ZBT fuel cell stack design from a mechanical point of view several experiments are performed and presented in this section. Several experimental analyses have been reported for other stack designs focusing on the internal stress distribution and the stack deflection [115, 117, 120-123]. The mechanical characteristics of the fuel cell stack and components are investigated utilizing diverse measurement methods as given in literature [121-123]. The effect of the mechanical clamping on the cell performance is investigated in [120]. The effect of the cell operation conditions on the mechanical properties of the fuel cell stack has not been properly examined in literature yet. In order to analyze the fuel cell stack from mechanical point of view under operating conditions, an experimental set-up is established. Several experiments are performed with this set-up to determine the thermomechanical characteristics of the fuel cell stack during operation. A stack with 5 cells is selected in order to examine the stack characteristics. The standard ZBT fuel cell stack design with 5 cells has been tested with a wide variety of different cell components. The thermomechanical properties of the standard ZBT fuel cell stack are presented in the first subsection. The performance and the operating points during these experiments are also given in this subsection.

The main aim of the mechanical clamping components is to assure the tightness of the stack and to deliver the required mechanical pressure for each cell and stack component as mentioned before. An even distributed mechanical pressure on the cell components plays a significant role for the fuel cell performance [52-54]. In order to analyze the mechanical stress distribution inside the fuel cell, pressure measurement films are used. The use of the pressure films and the measurements are presented in the second subsection. Experimental cell performance results of standard ZBT fuel cell stack are given in the last section.

3.1 *Measuring the Thermal Expansion of the Fuel Cell Stack*

The experimental set-up used to measure the thermal expansion of a stack with 5 cells is illustrated in Figure 3.1. The stack is assembled consciously with non-used components to be able to analyze the mechanical properties properly and to detect the influence of aging phenomena. The experimental set-up is built as depicted in Figure 3.2.

The medium suppliers are configured on one endplate of the stack to let the other side free for the thermal expansion measurements as seen in Figure 3.1 and Figure 3.2. The disc springs are assembled on the upper side of the fuel cell stack. The dial extensometer is preferred with a precision value of 1[µm] for the measurements due to its reliability and simplicity. The dial extensometer measures the change in the length of the fuel cell stack under operating conditions, which delivers the stack expansion values. The connection of the dial extensometer for logging of the data is established via USB port.

The dial extensometer is located in the middle of the endplate in order to get the whole thermal expansion values by preventing the influence of the buckling effect occurring via the tie rods. This promises stable measured data for each test.

In order to prevent the expansion in inverse direction, the fuel cell stack is placed on reference blocks as illustrated in Figure 3.1 and Figure 3.2. Two long blocks are settled under the fuel cell stack to achieve sufficient free space for the peripheral components without disturbing the experiments.

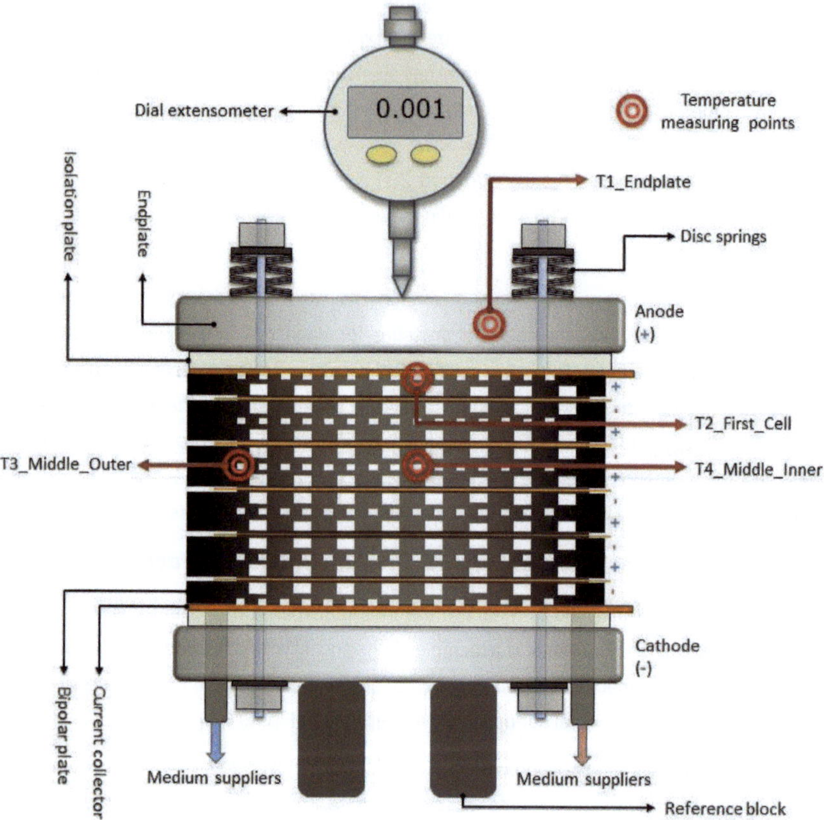

Figure 3.1 Illustration of the experimental set-up for the thermal expansion measurements

Thermal management of the fuel cell stack is accomplished with the use of a radial fan. The air cooled fuel cell stack is held under standard laboratory conditions.

The temperature distribution within fuel cell stacks is investigated for several purposes with different techniques [55-59].The most appropriate and stable measuring technique (regarding the ZBT fuel cell stack properties) is using thermocouples. The temperature data can be simultaneously logged with the expansion measurements, which facilitates the analysis of the thermomechanical characteristics of the fuel cell stack. The temperature measuring points and their positioning are precisely selected in accordance with the aim of thermal expansion measurements as seen in Figure 3.1. The thermocouples are installed in the air cooling channels of the bipolar plates and in a hole of the endplate in order to get the real temperature values of the components. The single cooling channels, in which the thermocouples are installed, are closed precisely to stabilize the measured data.

The temperature of the endplate (T1_Endplate) is a key value to analyze the thermal expansion characteristics of the fuel cell stack. The free convection and the remaining thermal energy passing through the isolation plate are determining the temperature of the endplate. Comparing the temperature of the endplate with the temperature of the first fuel cell provides also significant information to evaluate the function of the isolation plate. The temperature measuring point on the first cell from the top view (T2_First_Cell) is assumed to deliver the average temperature value of the regarding fuel cell due to the high thermal conductivity of the bipolar plate material. It is the last supplied fuel cell.

Figure 3.2 Experimental set-up for the thermal expansion measurements

The temperature measuring point (T3_Middle Outer) is located in the outer cooling channel of the middle fuel cell as seen in Figure 3.1. It delivers the temperature values of the middle fuel cell from the outermost air cooling channel. This enables to compare the temperature gradient along a single cell. The thermocouple (T4_Middle_Inner) provides the temperature value of the fuel cell positioned in the middle of the fuel cell stack. The thermocouples provide the analysis of the temperature distribution in the fuel cell stack and enable the correlation with the expansion measurements. The analysis of the thermal distribution assists to understand the temperature gradients within stack components. This corresponds to the examination of the thermal expansion load acting on the stack components.

3.1.1 Operating Conditions and Measurements

Appropriate operating points are to be chosen for a proper analysis, which represent the fuel cell stack behavior under real conditions. In order to analyze the dynamic thermomechanical behavior of the fuel cell stack, cyclic operating conditions are defined. The stack is operated until it reaches the full load conditions as given in Table 3.1.

Table 3.1 Full load operating conditions

I	20 [A]
T	70 [°C]
Cathode	$\lambda_{O_2} = 2$
Anode	$\phi_{RH} = 70\%, \qquad \lambda_{H_2} = 3$

A gradual increase in current drawn from the fuel cell stack by an average voltage value of 690[V] and waiting until it reaches the predefined maximal operating temperature is the first part of the cycle and called warming up process. The second part of the cycle is allowing the fuel cell stack to cool down with the help of a radial fan until it reaches the initial ambient temperature about 23°C. This part is called cooling down process and performed without medium supply. These operating points represent the real maximum conditions regarding the dynamic thermal expansion load on the fuel cell stack. The maximum and minimum thermal expansion values can be examined by these cyclic operating conditions. The thermal expansion values are recorded with the help of the dial extensometer during the cyclic tests with other measured data.

The temperature development in the fuel cell stack and the deflection of the stack length during the first cycle test can be seen in the Figure 3.3. The temperature in the fuel cell stack develops as expected. The resulting temperature difference between the endplate and the fuel cells is significant. It takes additional time for the thermal energy to reach the endplates. This is caused due to the use of isolation plates as explained before. The temperature gradient in the middle cell along the bipolar plate long side (T3_Middle_Outer and T4_Middle_Inner) can be remarked. This is caused by the effect of the supplementary convectional cooling on the outer surface of the bipolar plate and the reaction rate in the fuel cell. The temperature values on the first and middle fuel cell can also be noticed. The difference in the temperature values of the middle and first cells depends on several properties e.g. conductive heat transport on the last bipolar plate and media supply. The reaction of the thermal expansion to changes in temperature is quite fast as it can be seen in Figure 3.3. The increase and decrease in the temperature of the fuel cell stack cause a simultaneous change in the value of thermal expansion. The heat transfer rate to the endplate is reduced due to the use of the isolation plate. This results in a time delay for the thermal response of the endplate. The stack deflection reaches an approximate value of 94[μm] when the stack reaches the full load operating point as given in Table 3.1. The stack deflection value is stabilized, when the stack temperature values stay constant.

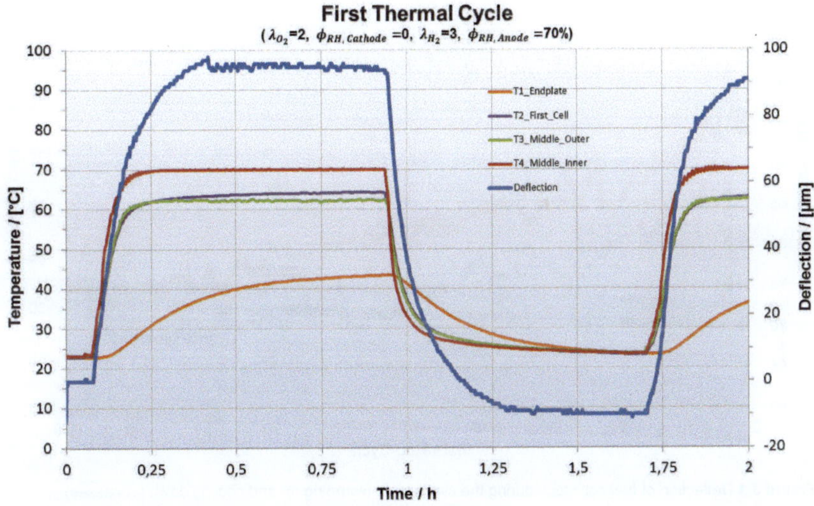

Figure 3.3 Deflection and temperature values during the first thermal cycle

The first cycle test introduces the stack deflection and the operating conditions to be examined, which could affect the stack deflection. The stack shows a negative deflection at the end of the cooling down process. The shrinkage of the fuel cell stack occurs due to the plastic deformation of cell components, which denotes the irreversible deflection occurring in the material. Elongation via thermal expansion and compression of the fuel cell stack induce the plastic deformation. The mechanical stress acting on the cell components increases with the thermal stress caused by the warming up process. The stack components e.g. GDL and silicone seals exhibit strong inelastic deformation behavior. The disc springs balance the stack deflection to assure the identical clamping force on the fuel cell stack. This results in a permanent reduction of the total fuel cell stack length. This phenomenon can be precisely observed by the first cycle test given in Figure 3.3. The shrinkage of the fuel cell stack for the first cycle test has an approximate value of -10[μm].

Therefore, particularly the disc springs are to be concerned for a proper mechanical design. The disc springs are selected and configured according to the spring characteristic curve (force-deflection curve) regarding the desired axial force applying on the stack. It is crucial to achieve a flat spring characteristic curve to provide the desired axial compression force for the fuel cell stack. Flat spring characteristics lead to the identical axial force acting on the fuel cell stack independent from the stack deflection.

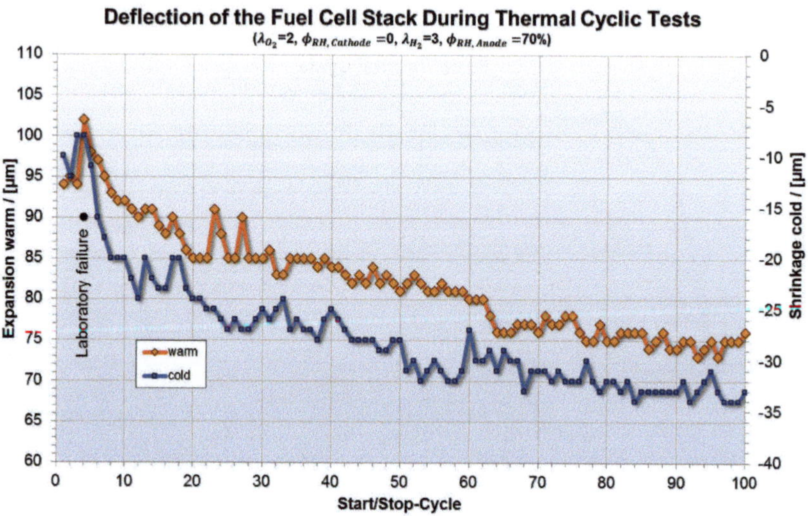

Figure 3.4 Deflection of fuel cell stack during the cyclic tests (warming up and cooling down processes)

The expansion and shrinkage of the fuel cell stack and their properties depending on the number of driven cycles are to be examined further for a better understanding. For this purpose the cyclic tests are performed further with the same fuel cell stack. The stack is operated until it reaches the full load conditions as defined before and then it is released to cool down until the ambient temperature. The measured data during thermal cycling processes as defined before is given in Figure 3.4. The red line in Figure 3.4 exhibits the values of the thermal expansion peaks by warming up process for each cyclic test. The blue line shows the values of the fuel cell stack shrinkage at the end of the cooling down process for each cycle. As it can be extracted from the trend of the both lines in Figure 3.4, the difference between the deflection peaks on the fuel cell stack are moderately higher during first cyclic tests. The more the stack is used, the more stabilized thermomechanical characteristics are measured for the fuel cell stack. The plasticity of the stack components decreases due to the cyclic operation conditions. This phenomenon is dealt with the definition of the material properties of cell components e.g. gas diffusion layers and sealings given in section 4.2.2.4. The thermal expansion of the fuel cell stack stays approximately constant after 60 cycles at a value of 77[μm], the shrinkage at a value of -30[μm].

3.1.1.1 The Effect of the Fuel Cell Operating Parameters on the Deflection

In Figure 3.5 the operating parameters and measured data in the first thermal cyclic test are given. As expected it can be extracted from Figure 3.5 that the deflection of the fuel cell stack tends to respond directly to the temperature changes in the fuel cell stack but not to the current and voltage values. This can be noticed by following the deflection rate of the fuel cell stack after sudden changes in the current and average cell voltage values.

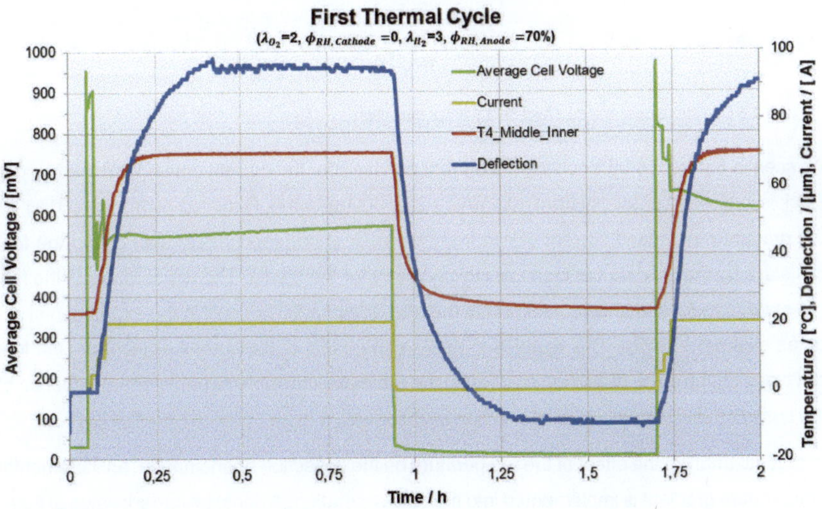

Figure 3.5 Operating parameters and deflection of the stack during the first thermal cycle

Additional supplementary tests are necessary in order to understand the effect of the single operating parameters on the deflection of the fuel cell stack. The selected operating parameters and their effect on the deflection of the fuel cell stack can be followed simultaneously by setting other operating parameters constant. Thus the effect of the corresponding operation parameter on the deflection of the fuel cell stack can be analyzed properly.

In Figure 3.6 the current density is varied incrementally in order to investigate the effect of the current density changes on the stack deflection.

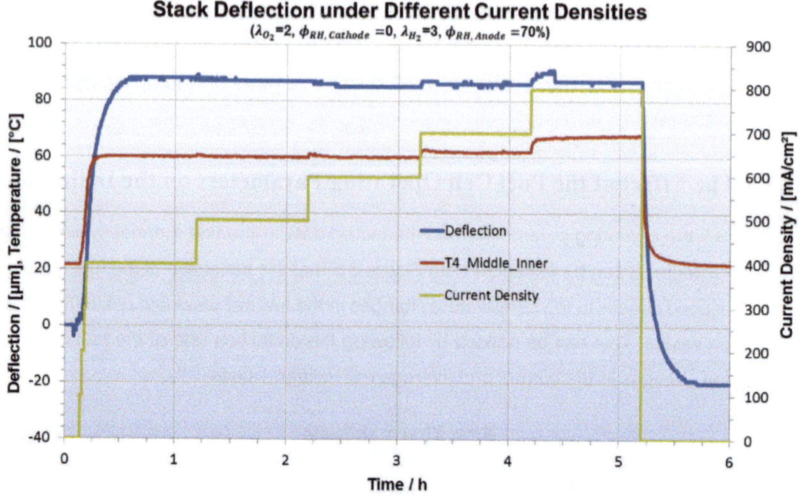

Figure 3.6 Current density and its effect on the stack deflection

As seen in Figure 3.6 by following the current density line, the current drawn from the fuel cell stack is increased gradually from 400[mA/cm2] to 800[mA/cm²]. Required hydrogen and air for the gradually increased current drawn from the stack is correspondingly supplied to the fuel cell stack by maintaining the identical stoichiometry for the media. The temperature of the fuel cell stack is regulated to be constant with the help of the radial fan built in the experimental set-up as explained before. This enables to analyze the effect of the current drawn from the fuel cell stack on the stack deflection discarding the thermal expansion. No substantial influence of the current density on the stack deflection is observed.

In order to analyze the effect of the temperature on the deflection phenomenon, an incremental temperature gradient is implemented into the stack operation. A constant current value of 20[A] is drawn from the fuel cell stack by a constant voltage value about 690[mV]. The fuel cell stack is operated with identical hydrogen and air supply as cyclic testing conditions given in Table 3.1. The gradual increase and decrease in temperature profile are provided by regulating the radial fan depending on the temperature value of the cell positioned in the middle of fuel cell stack (T4_Middle_Inner). Stepwise temperature gradients developed on the fuel cell stack are given in Figure 3.7 for the whole temperature measurement points.

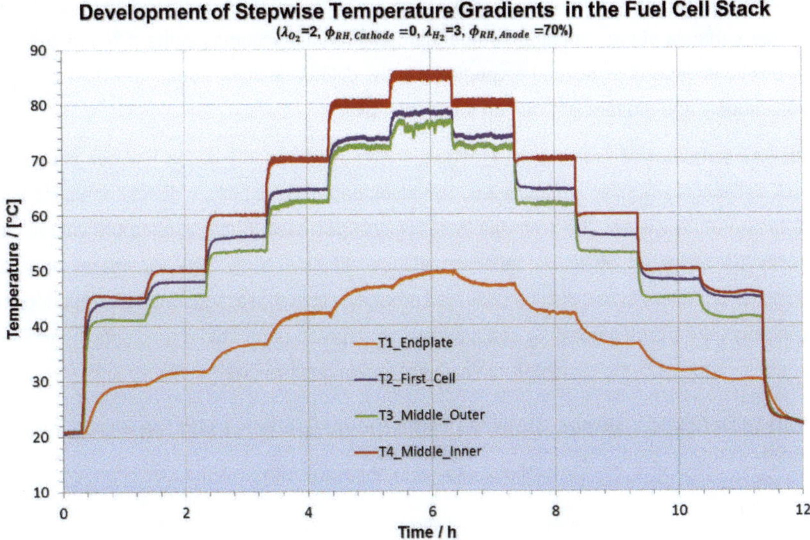

Figure 3.7 Development of stepwise temperature gradients at 4 different positions in the fuel cell stack

Temperature gradients occurring in the fuel cells reach endplate with a time delay as it can be seen in Figure 3.7. Thermal response of the endplate is formed by the use of components particularly isolation plate as explained before.

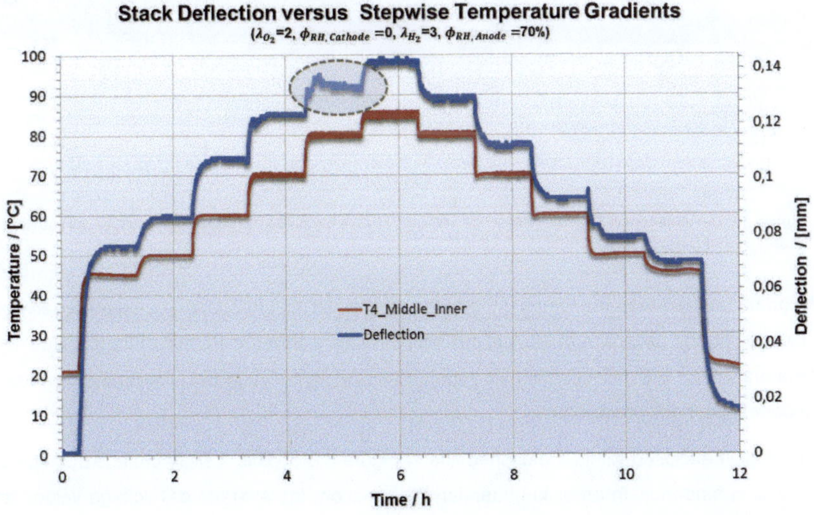

Figure 3.8 Development of the stack deflection versus stepwise increased temperature in the fuel cell stack

The temperature development at the endplate is also affected by free convection occurring on the free surfaces of the endplate. Temperature differences occurring in the different cell and cell positions depend on several properties e.g. the difference in the conductive heat transfer, media supply and the reaction rate on the bipolar plate.

Stepwise temperature gradients (T4_Middle_Inner) in the middle of the fuel cell stack and stack deflection are given in Figure 3.8. The stack deflection exhibits a direct correlation with the temperature values. The stepwise increase and decrease in the temperature values can directly be observed by stack deflection values simultaneously. The expansion and the shrinkage phenomena can also be observed properly within these diagrams by comparing the deflection values during upwards and downwards stepwise temperature gradients. In addition to data in Figure 3.8, the temperature values of anode gas are given in Figure 3.9.

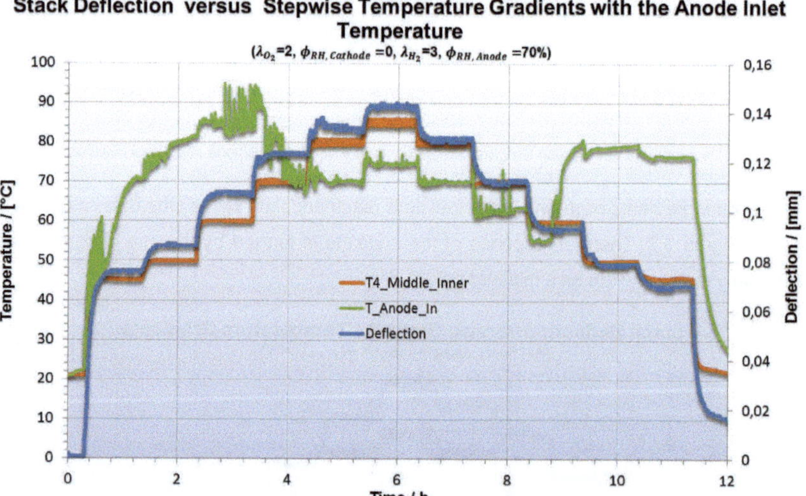

Figure 3.9 Development of the stack deflection versus stepwise temperature gradients in the fuel cell stack with anode inlet temperature

It can be realized from the Figure 3.9 that the anode gas inlet temperature has no observable influence on the deflection of the fuel cell stack. In this manner, the effect of the cathode inlet temperature on the deflection is also regarded as negligible regarding the air supply mostly under atmospheric conditions.

The region enclosed within the dashed line in Figure 3.8 depicts a slight change in stepwise deflection tendency. In order to understand the reason, the average cell voltage values are also given additionally together with the stepwise temperature values in Figure 3.10. As mentioned before a constant current value of 20[A] is drawn from the fuel cell stack during gradual temperature increase and decrease. Performance of the fuel cell is decreasing after fuel cell stack passes over the operation temperature value of 70[°C] due to the cell voltage drop as it can be seen in Figure 3.10.

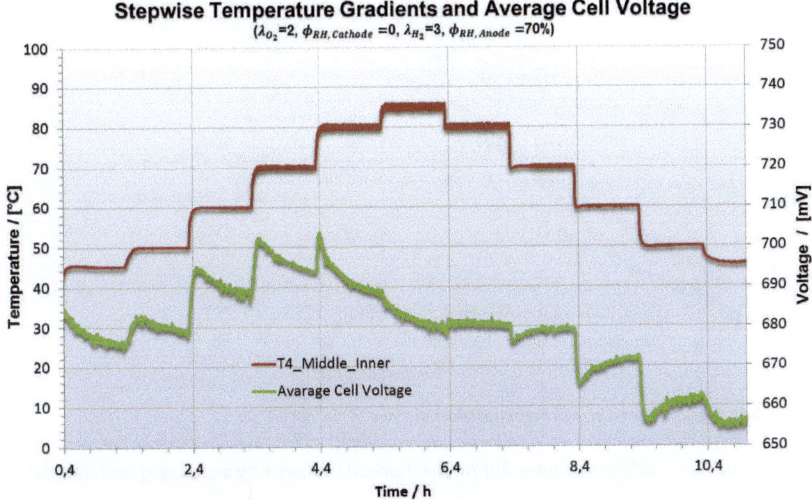

Figure 3.10 Stepwise temperature gradients in the fuel cell stack and average cell voltage

Average cell voltage drop occurs during the stepwise temperature gradient after about 4 hours, which can be evaluated together with the stack deflection in the enclosed region within the dashed line in Figure 3.8. In other words the temperature of the fuel cell stack is so increased, which results in drying out of the membrane and affects directly cell performance [60-62]. The membrane swells when it is humidified [63-65]. Hence the dehydration of membrane induced by the temperature increase causes a shrinkage of the membrane. This phenomenon can be observed during these measurements by following the temperature gradients and the average voltage values in Figure 3.10 together with the stack deflection values in Figure 3.8. A higher number of fuel cells in the stack increases the influence of this phenomenon on the stack design. The difference in the average cell voltage values during upwards and downwards stepwise temperature gradients occurs due to the regulation of the humidification level of membrane. In order to analyze this slight decrease in the stack deflection within the enclosed region in Figure 3.8, the stack deflection versus temperature (T4_Middle_Inner) is plotted in Figure 3.11. The stack deflection is given in a function of stack temperature instead of time. This enables to follow the stack deflection values and the expansion behavior of the fuel cell stack depending on the change in the cell temperature. The dashed red arrow defines the warming up process direction and the green one is for the cooling down process direction.

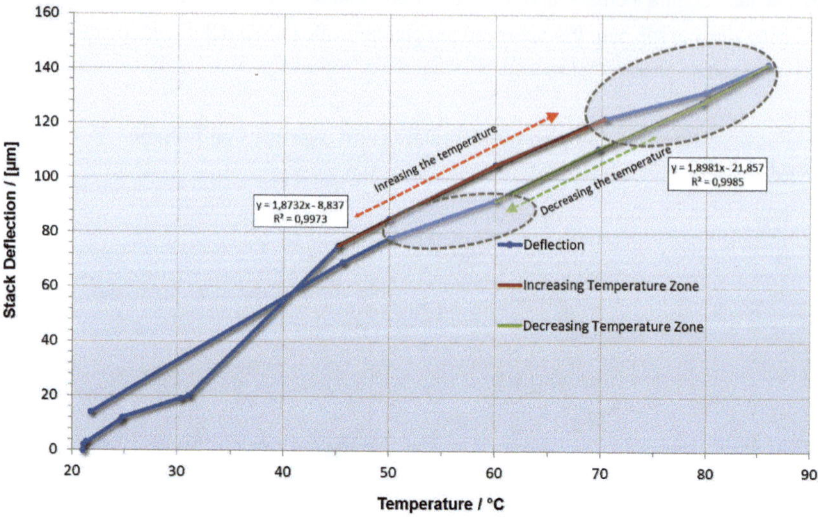

Figure 3.11 Stack deflection versus the temperature gradient in the fuel cell stack

The fuel cell stack exhibits a linear thermomechanical behavior by increasing and decreasing temperature values as it can be seen in Figure 3.11. The thermal expansion coefficient for the whole fuel cell stack could also be extracted by using the data plotted in Figure 3.11. During heating of the fuel cell stack the drying out effect of the membrane takes place above 70°C, which is also marked in the region within the dashed lines above. The marked region within the dashed lines below defines the membrane humidification. The linear thermomechanical behavior of the fuel cell stack is slightly interrupted due to the membrane swelling factor by warming up and cooling down processes. This effect could be observed precisely with the help of Figure 3.11. The stack deflection values at the beginning of the warming up process and the end of the cooling down process differ from each other slightly due to the swelling of the membrane caused by the different operating conditions.

Mechanical properties of the standard ZBT stack with 5 fuel cells are measured with the help of an experimental set-up during cell operation and analyzed depending on several operating parameters and cyclic conditions. Total deflection of standard ZBT stack with 5 fuel cells has an approximate value of 94[µm] at the end of the expansion process for the first cycle. The fuel cell stack exhibits also a shrinkage behavior and has a deflection value of about -10[µm] for the first cycle, which is induced by the plastic deformation of the stack components. Mechanical behavior of the fuel cell stack stabilizes after about 60 cycles and hasn't changed significantly anymore. Total deflection values of the fuel cell stack after 60 driven cycles are about 77[µm] and-30[µm] for thermal expansion and shrinkage process respectively as it can be extracted from Figure 3.4. Swelling of the membrane due to the membrane humidification is also observed and has an average value of about 6[μm] for the stack with 5 fuel cells. Comparing the size of the (de)hydration effect with the total stack deformation and considering the dependency only on the (de)hydrated cases of the membrane, this phenomenon is not regarded in the simulations performed in section 4. The deflection measurements can be extended with the use of optical measuring devices or additional dial extensometers.

3.2 Pressure Films

An even distributed mechanical pressure on the cell components plays a significant role for the fuel cell performance [52-54]. Pressure measurement films are used for the analysis of the mechanical pressure distribution inside the fuel cell [115, 117, 121, 122]. Digital pressure sensors cannot be utilized for a standard ZBT stack with 5 cells due to the size and geometry of the available pressure sensors, which restricts the application of the tie rods. Digital pressure sensors can be applied for standard ZBT stack design with more than 10 fuel cells by kinking the pressure sensors without disturbing the tie rods. The [a]Prescale® pressure measurement films are utilized to visualize the distribution of the mechanical load in standard ZBT fuel cell stack. The ultra-thin Prescale® film enables the application between the cell components. The position of the pressure measurement film in the fuel cell stack is illustrated in Figure 3.12.

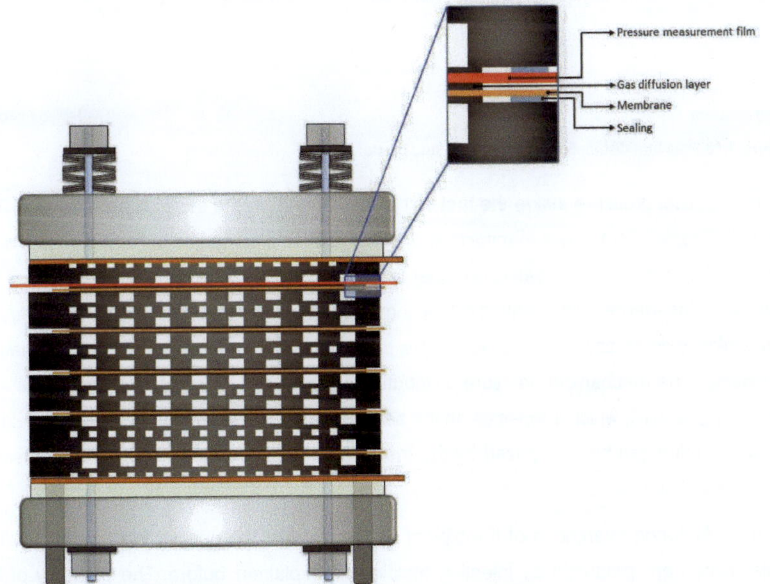

Figure 3.12 Position of the pressure measurement film in the fuel cell stack

---

[a] Trademark of Fujifilm Corp.

Sealing and gas diffusion layer (GDL) are standard components and have approximate thickness values of 230[µm] and 190[µm] respectively. An additional gas diffusion layer is used in order to get sufficient mechanical pressure on the pressure measurement film as it can be seen in Figure 3.12.

Application temperature of the pressure measurement films is between 20°C and 35°C, which restricts their use in fuel cells under operating conditions [66]. However they provide a better understanding of the mechanical pressure distribution inside the fuel cell. The pressure distribution on the pressure measurement film is given in Figure 3.13.

Figure 3.13 Pressure distribution between GDL and bipolar plate

The mechanical pressure inside the fuel cell is the highest on the sealings as expected. The form of the sealing for the gas manifolds and fuel cell active area can be noticed in Figure 3.13. The thickness of the silicon sealing is higher than the gas diffusion layer in order to ensure the tightness of the fuel cell. The sealing is first compressed until the thickness of the gas diffusion layer, which is compressible and porous. This results in higher mechanical pressure values on the sealing. The mechanical pressure distribution on the membrane active layer can be also seen in Figure 3.13, which intensifies on the sealing borders. The structure of the flow field on the bipolar plate can be recognized by the mechanical pressure distribution on the pressure measurement film.

The manufacturing tolerances of the bipolar plates can also be noticed in Figure 3.13. The bipolar plates are produced by injection molding as explained before. The planarity of the bipolar plate is not sufficient due to the manufacturing processes by injection molding. The surfaces of the bipolar plate are milled additionally, in order to improve the surface roughness and the planarity of the plate. The milling paths on the surfaces of the bipolar plate can be remarked by the measurement of the mechanical pressure distribution as it can be seen in Figure 3.13.

Exact values of the mechanical pressure are not available by the performed measurements with the help of pressure measurement films but the mechanical pressure distribution inside the fuel cell is provided with the help of experiments, which is utilized for the verification of the simulations performed in section 4.2. The geometrical and manufacturing tolerances of fuel cell stack components e.g. bipolar plates are not taken into consideration by the simulations, which has to be regarded by comparing the measured data with the simulations results.

3.3 Cell Performance

The performance of a fuel cell stack can be evaluated with the use of polarization curve, which is a typical diagnostic diagram of the cell voltage versus current density, derived by varying the current drawn from the fuel cell stack as explained before.

Standard ZBT stack with 5 fuel cells and an active area of 50cm² is operated at 67°C and the polarization curve is generated by varying the drawn current from the fuel cell stack between 0-50 [A], which is given in Figure 3.14.

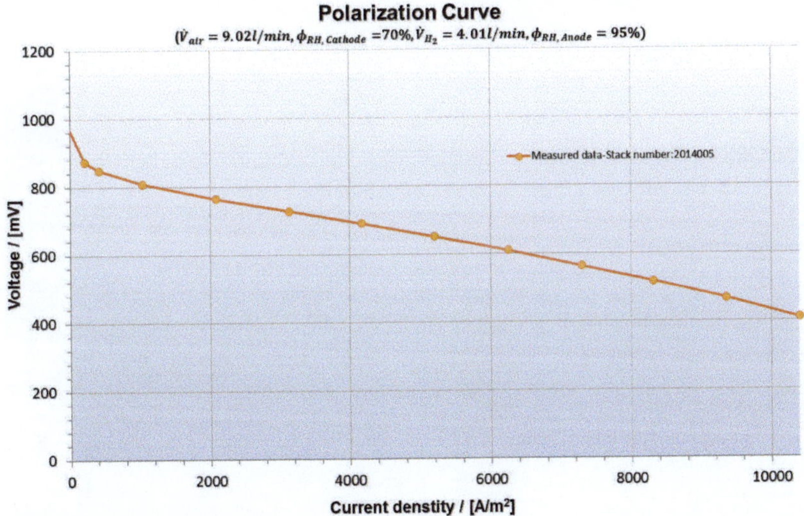

Figure 3.14 Polarization curve of ZBT Stack (Stack Number:2014005)

The average voltage values of fuel cells for different current values form the polarization curve plotted in Figure 3.14. Fuel cell stack is supplied with constant media with about 4.01[l/min] hydrogen and 9.02[l/min] air by generation of the polarization curve. This enables the utilization of the measured data for the verification of the simulations performed in section 4.3. The measured medium flow rates by generating the polarization curve plotted above are implemented into the electrochemical model to define the boundary settings for the fluid dynamics. This enables the utilization of the measured polarization curve to verify the results of the electrochemical simulations performed in section 4.3.

4 Fuel Cell Modeling

There has been tremendous progress in computational methods due to the development in numerical capabilities. Computational methods have become standard tools for analysis during design processes [67].

The use of computational methods for fuel cell technology facilitates the analysis during the prototyping process and enables to get immeasurable data due to the spatial dimensions of the fuel cell components.

Fuel cell models built with the help of programming platforms like [b]MATLAB/Simulink®, [c]Dymola/[d]Modelica, [e]Scilab etc. usually base on a mathematical approach and don't include the geometrical and physical properties [68-71]. These mathematical models are used both for fuel cell analysis and control systems. Detailed information about the modeling of fuel cells with the help of numerical platforms can be found in [72-78].

Geometry based analysis of fuel cells with the help of numerical discretization is another method to study fuel cells in detail. Numerical discretization methods are based on the definition of the mathematical models in terms of partial differential equations for a given domain. The linearization of the mathematical models is achieved via the discretization of the models in the corresponding domain. This leads to numerically solvable models with the help of computational algorithms.

The finite difference, finite volume and finite element methods are common numerical discretization methods developed and used for geometry based modeling. The mathematical approach for the linearization of the mathematical models differs in each method. The difference between the mathematical backgrounds and the functionality of each method are handled in detail in [79-84]. There are several available software packages for each discretization method. The survey for the most appropriate software package is performed precisely regarding the point of interest of this work. The multiphysics capability is required to simulate the interaction of several physical properties, which regularly occurs in the field of fuel cells. [f]COMSOL Multiphysics® is chosen as simulation platform for the computations performed in this section because of its suitable functionality and multiphysics capabilities. COMSOL Multiphysics® is based on the finite element method. The basic introduction of the finite element analysis is given in the following subsection for a better understanding of the computations. Two simulation models are presented in section 4.2 and 4.3 respectively, which are

---

[b] Trademark of MathWorks, Inc.
[c] Trademark of Dassault Systèmes
[d] Trademark of Modelica Association
[e] Trademark of Inria
[f] Trademark of COMSOL AB

established with the help of COMSOL Multiphysics® and are handling the standard ZBT fuel cell stack from a mechanical point of view within the framework of this study. The developed simulation models are illustrated in Figure 4.1 together, in order to provide a proper understanding.

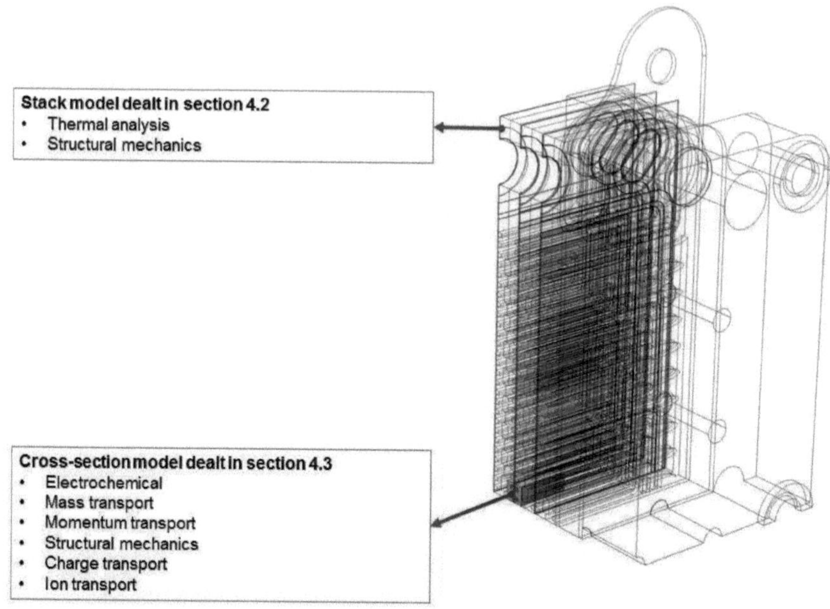

Figure 4.1 Illustration of stack and cross-section model

A first model (wireframe geometry in Figure 4.1) is developed to analyze the thermal expansion and mechanical stress distribution on the standard ZBT stack with 5 fuel cells. As introduced in the section 3 the fuel cell operation conditions don't intensely affect the mechanical properties of the fuel cell stack excluding the temperature. Hence internal stress distribution of the fuel cell stack and its temperature dependency are studied in section 4.2. The mechanical stress distribution on each stack component can be investigated with the help of this model, supplemental to the measurements presented in section 3. This provides immeasurable data due to the experimental restrictions and expands the evaluation of the analysis from single measuring points to the whole computational domain.

Electrochemical modeling of a fuel cell stack with 5 fuel cells containing structural mechanics and fluid dynamics is not applicable with the help of the numerical discretization methods regarding available computational capacity. The second model (highlighted geometry in Figure 4.1) dealt in section 4.3 is established by taking a cross-section of a single fuel cell considering the electrochemistry. Fuel cell parameters are studied in dependence of the compression characteristics. Stress distribution on the gas diffusion layers, which is computed in the stack model, is utilized for the evaluation of the cross-section model.

4.1 Introduction to Finite Element Method

Finite element analysis is a numerical method to get approximate solutions of the differential equations. This section introduces the basic approach to develop the approximate solution of boundary value problems for differential equations. This leads eventually to the implementation of the numerical algorithms.

The mathematical problem definition is established with the help of differential equations and initial conditions for a given computational domain (e.g. Figure 4.2). This computational domain is splitted into tiny parts which denotes the term of "finite element" (see Figure 4.3). This procedure is called meshing process. The assembly of the finite elements represent the whole computational domain. The approximation of the explicit solution via the predefined functions for each element corresponds to the linearization. The discretized mathematical problem is solved for each finite element with the help of boundary values given in the problem definition. Due to the reassembly of all finite elements, the solution for each element leads to the approximate solution for the whole domain.

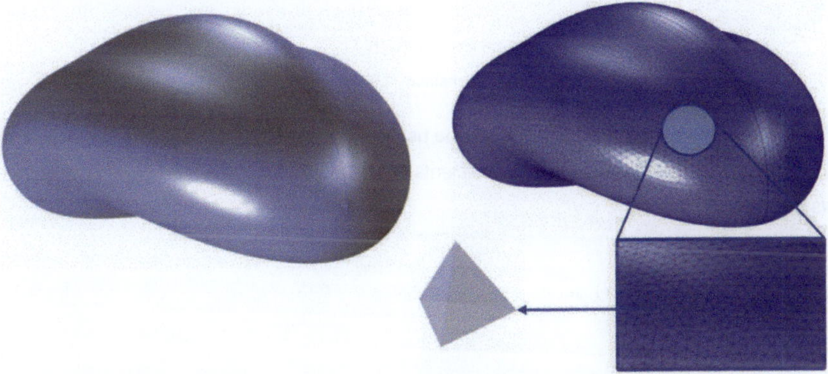

Figure 4.2 3-dimensional computational domain   Figure 4.3 Meshing of the corresponding computational domain and tetrahedral element

The accuracy of the approximated solution can be improved by intensifying the meshing properties, which increases the required computational capacity. The outline of the finite element analysis is pointed out in Figure 4.4 for a better understanding of the mathematical background.

The problem definition for given initial and boundary conditions is the starting point of finite element method as seen in Figure 4.4. The approximation of the explicit solution $u(\vec{x})$ (see Eq. 4.1) is established with the implementation of $\psi_a(\vec{x})$ called shape function. N is the number of the elements splitting the corresponding domain.

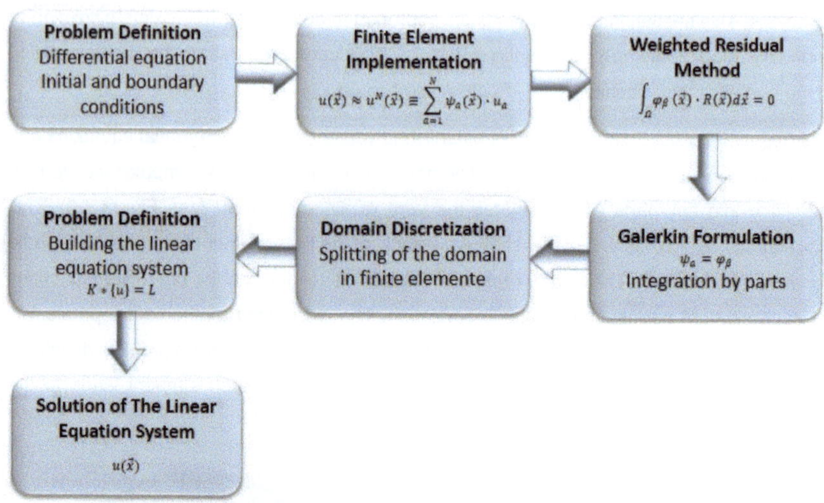

Figure 4.4 The basic outline of the finite element method

Polynomials or Lagrange formulation can be used as shape function [85, 86]. In order to find the best approximated values of the coefficients u_a Galerkin formulation (see Eq. 4.2) is utilized as depicted in Figure 4.4.

$$u(\vec{x}) \approx u^N(\vec{x}) = \sum_{a=1}^{N} \psi_a(\vec{x}) \cdot u_a \qquad\qquad \text{Eq. 4.1}$$

$$\int_\Omega \varphi_\beta(\vec{x}) \cdot R(\vec{x})d\vec{x} = 0 \qquad\qquad \text{Eq. 4.2}$$

Galerkin formulation is based on the problem definition to establish the residual function $R(\vec{x})$ given in Eq. 4.2. Multiplying the residual function with arbitrary weighting function $\varphi_\beta(\vec{x})$ and integrating the residual expression on the corresponding domain lead to the Galerkin formulation. The integral expression is forced to vanish for the purpose of approximation. Choosing the weighting function $\varphi_\beta(\vec{x})$ same as shape function $\psi_a(\vec{x})$ reduces the problem definition.

The principle of integration by parts is utilized to reduce the second order derivative term. This process is called as weak formulation from mathematical point of view.

Implementation of the approximation function $u(\vec{x})$ in the weak expression corresponds to a linear equation system for unknown u_a given in Eq. 4.3.

$$K * \{u\} = L$$

Eq. 4.3

K and L are stiffness matrix and load vector respectively. u is the solution vector formed by the unknown values of the knots for each finite element, which represent the coefficients of the approximated expression (see Eq. 4.1). The stiffness matrix is a large sparse matrix depending on the number of the elements splitting the corresponding domain. The load matrix is formed by the boundary conditions given in the problem definition. The solution of the linear equation system for unknown u leads to the approximated solution of the boundary value problem.

In order to understand the finite element method properly, a simple one dimensional example is taken into account in the following. The solution of the simple boundary value problem is given with the help of finite element method step by step [85-90]. The results are compared with the explicit solution, which yields a better understanding of the finite element method. An elliptic ordinary differential equation form is taken as example, which appears in several physical phenomena with the defined boundary and initial conditions given in Eq. 4.4 and 4.5 [85-90].

$$\frac{d^2 f}{dx^2} = -1 \ , x \in \Omega := [0,1] \subset \mathbb{R}$$

Eq. 4.4

The boundary conditions for this problem definition are:

$$f(0) = 0, \qquad f(1) = 0$$

Eq. 4.5

As explained before this boundary value expression is multiplied with an arbitrary weighting function $\varphi: \Omega \to \mathbb{R}$ with $\varphi(x) = 0, x \in \Omega$ and forced into integral for an interval of $\Omega := [0,1] \subset \mathbb{R}$ analogous to Eq. 4.2 as seen in the Eq. 4.6.

$$\int_\Omega \left(\frac{d^2 f}{dx^2} + 1 \right) \varphi dx = 0$$

Eq. 4.6

The principle of integration by parts is applied to the Eq. 4.6 to reduce the order of the derivation as in Eq. 4.7.

$$\int_\Omega \frac{d^2 f}{dx^2} \varphi dx = \left[\frac{df}{dx} \varphi \right]_0^1 - \int_\Omega \frac{df}{dx} \frac{d\varphi}{dx} dx = - \int_\Omega \frac{df}{dx} \frac{d\varphi}{dx} dx$$

Eq. 4.7

The integral term vanishes and the following weak expression occurs from Eq. 4.6 and Eq. 4.7.

$$\int_\Omega \frac{df}{dx}\frac{d\varphi}{dx}\,dx = \int_\Omega \varphi dx$$

Eq. 4.8

The domain Ω is divided into a number of $N = 4$ intervals $\tilde{\theta}_i$ with the equal element size h. In order to calculate the integration part in Eq. 4.8, the following two approximation functions (see Eq. 4.9-10) are defined with the help of chosen shape functions $\psi_i(x)$. The shape functions are defined specially for each interval $\tilde{\theta}_i$.

$$u_a = \sum_{i=1}^{N} \psi_i(x)\cdot u_i$$

Eq. 4.9

$$\varphi_a = \sum_{i=1}^{N} \psi_i(x)\cdot \varphi_i$$

Eq. 4.10

The shape functions are special functions chosen appropriately for the purpose of the finite element method. In this example linear shape functions are taken into consideration, which have a value of zero out of each corresponding interval $\theta_i \subset \Omega$. The interval $\tilde{\theta}_i := [x_i, x_{i+1}]$ denotes the non-zero region for each corresponding shape function as seen in Eq. 4.11.

$$\psi_i(x) = \begin{cases} \dfrac{x - x_{i-1}}{h} & x \in \tilde{\theta}_{i-1} \\ \dfrac{x_{i+1} - x}{h} & x \in \tilde{\theta}_i \\ 0 & else \end{cases}$$

Eq. 4.11

The graphical representation of shape functions through the splitted computational domain is depicted in Figure 4.5 to gain a better understanding.

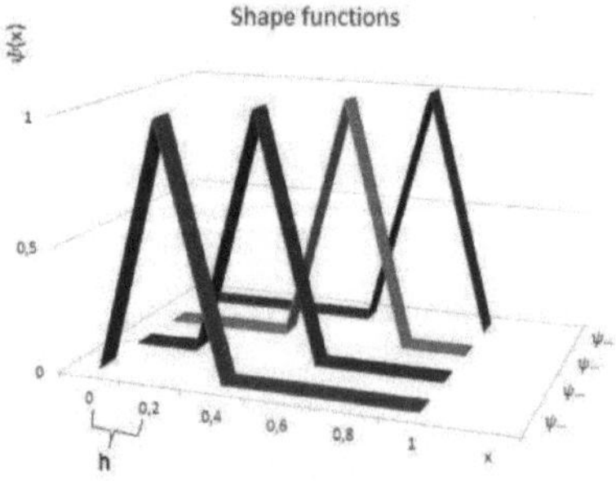

Figure 4.5 Shape functions illustrated on the computational domain

As explained before the domain Ω is divided into a number of $N = 4$ intervals $\tilde{\theta}_i$ with equal element size of h as depicted in Figure 4.5. Multiplying each linear shape function with the value of corresponding element and the summation of these expressions gives the approximation equation Eq. 4.9. The derivation of each shape equation can also simply taken into account due to their linearity as given in Eq. 4.12.

$$\frac{d\psi_i(x)}{dx} = \begin{cases} \dfrac{1}{h} & x \in \tilde{\theta}_{i-1} \\ \dfrac{-1}{h} & x \in \tilde{\theta}_i \\ 0 & else \end{cases} \qquad \text{Eq. 4.12}$$

Implementing the approximated functions Eq. 4.9 and Eq. 4.10 in the weak residual expression Eq. 4.8 gives the following expression:

$$\sum_{ij} \varphi_i u_j \int_0^1 \frac{d\psi_i(x)}{dx}\frac{d\psi_j(x)}{dx} dx = \sum_i \varphi_i \int_0^1 \psi_i(x)dx \qquad \text{Eq. 4.13}$$

Using Eq. 4.12 the integral term in Eq. 4.13 can be calculated as in Eq. 4.14.

$$\int_0^1 \frac{d\psi_i(x)}{dx}\frac{d\psi_j(x)}{dx} dx = \begin{cases} \dfrac{2}{h} & i = 1 \\ -\dfrac{1}{h} & |i - j| = 1 \\ 0 & else \end{cases} \qquad \text{Eq. 4.14}$$

The integral term in Eq. 4.13 on the right side is given in Eq. 4.15.

$$\int_0^1 \psi_i(x)dx = h \qquad \text{Eq. 4.15}$$

Implementation of Eq. 4.14 and Eq. 4.15 in Eq. 4.13 corresponds to a linear equation system analogous to Eq. 4.3 and is given in Eq. 4.16 for the given number of $N = 4$ shape functions.

$$h\begin{bmatrix} 1 \\ 1 \\ 1 \\ 1 \end{bmatrix} = \frac{1}{h}\begin{bmatrix} 2 & -1 & 0 & 0 \\ -1 & 2 & -1 & 0 \\ 0 & -1 & 2 & -1 \\ 0 & 0 & -1 & 2 \end{bmatrix}\begin{bmatrix} u_1 \\ u_2 \\ u_3 \\ u_4 \end{bmatrix} \qquad \text{Eq. 4.16}$$

This linear equation system can be solved simply and the solution vector u is as follows:

$$\begin{bmatrix} u_1 \\ u_2 \\ u_3 \\ u_4 \end{bmatrix} = \begin{bmatrix} 2h^2 \\ 3h^2 \\ 3h^2 \\ 2h^2 \end{bmatrix} \qquad \text{Eq. 4.17}$$

After implementation of the unknown values and shape functions into the Eq. 4.9, the approximation function takes the form as follows:

$$u_a = h^2 [2\psi_1(x) + 3\psi_2(x) + 3\psi_3(x) + 2\psi_4(x)]$$

Eq. 4.18

Eventually the analytical explicit solution of the boundary value problem and its approximation u_a derived by the finite element approach are both illustrated in Figure 4.6.

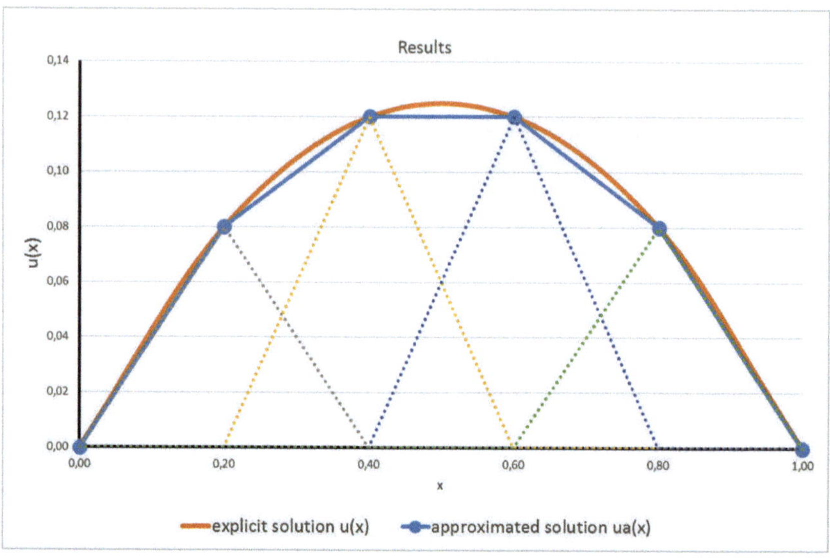

Figure 4.6 Explicit and approximated solutions of the boundary value problem

As it can be extracted from Figure 4.6, getting the approximated solution closer to the explicit solution improves the approximation procedure. Increasing the number of the elements splitting the computational domain Ω improves the approximation. Other type of shape functions can also be utilized to improve the approximation. This approximation method opens the gateway to the computational algorithms for the solution of the boundary value problems. The basic approach to the solution of the boundary value problems is same for all numerical discretization techniques.

The solution of the 3-dimensional boundary value problems with the help of finite element method has also the same principles as one-dimensional computational domains. The solution of the 3-dimensional domains results in the increase of the number of elements and the complexity of the stiffness matrix. Several tuning options are also developed for a better convergence depending on the definition of the boundary value problem [91-93]. COMSOL

Multiphysics® provides several shape function algorithms and enables the import of the computational domains built with other CAD platforms.

Further information about finite element method is given in detail in [94-109]. Computational methods for the multiphysics problems and application of the finite element approach for different physical phenomena are handled in detail in [110-112].

4.2 Fuel Cell Stack Simulations (Stack-Size)

A 3-dimensional thermomechanical model has been set up and solved using COMSOL Multiphysics®, in order to analyze the design of the fuel cell stack and the effect of the clamping on the single cell components. This provides supplementary information to the measurements presented in section 3 for the analysis of the mechanical characteristics of the fuel cell stack.

First a detailed literature survey of the fuel cell stack analysis from a mechanical point of view is performed and given in this section. Considering the point of interest and the state of the current technical approach, a 3-dimensional fuel cell stack geometry is prepared for the simulations concerning the computational properties and available capacity. A stack with 5 fuel cells is assigned for the simulations in this section analogous to the experimentations made in section 3.1 allowing a validation of the performed computations.

The development of the computational model is given in next steps. Thermal and structural properties of the model are defined in 4.2.2. Fatigue analysis is not considered by the modeling.

The results of the structural and thermomechanical simulations are given in section 4.2.3 separately in order to analyze the effect of the thermal expansion on the mechanics of the fuel cell stack.

4.2.1 Literature Model Overview

2-dimensional structural analysis of the fuel cells is investigated without electrochemical modeling in [113, 114]. Several 1 and 2-dimensional models are presented intensifying on the GDL compression properties and the effect of the compression on the fuel cell performance. These studies regarding the cell performance are examined in detail in section 4.3.1.

3-dimensional models have been recently developed for PEM fuel cell stacks with varying modeling properties. 3-dimensional studies investigating the mechanical properties of PEM fuel cell stacks don't take the thermal properties into consideration [115-117]. The geometrical details are typically simplified to reduce the complexity of the simulations for a 3-dimensional computational model.

The thermal expansion phenomenon of the fuel cell stack is a known parameter affecting the fuel cell stack mechanics. The effect of the thermal expansion varies depending on the type of the analyzed fuel cell. For high temperature PEM fuel cell stacks the thermal expansion values of the fuel cell stacks are getting higher. The thermal expansion of a high temperature PEM fuel cell stack with one single fuel cell is investigated in [118]. The thermomechanical analysis of a low temperature fuel cell stack with the help of computational methods has not been published yet.

The 3-dimensional model of the standard ZBT fuel cell stack is built containing the geometrical features in detail focusing on the thermomechanical characteristics.

4.2.2 Model Development

The modeling and the results of the thermomechanical simulations of a standard ZBT PEM fuel cell stack performed with COMSOL Multiphysics® version 3.5a are published in [119]. The presented FEM model in this study is built analogous to that model and performed with COMSOL Multiphysics® version 4.3b.

In this section the simulation model built with COMSOL Multiphysics® is described in detail in subsections. The preparation of the geometry and building of the 3-dimensional model is explained in subsection 4.2.2.1. The governing equations acting in corresponding subdomains are described in section 4.2.2.2. Then the boundary settings are given for the finite element model with the help of the problem definition. The material properties of the fuel cell stack components and meshing properties with the given solver settings are explained in subsections 4.2.2.4 and 4.2.2.5 respectively.

4.2.2.1 Model Set-up

Figure 2.11 is an exploded view of the 3-dimensional fuel cell stack model, which is derived with the help of the CAD platforms. This 3-dimensional model contains the detailed geometrical features for manufacturing purposes, which is also utilized to develop the FEM model in the following. The FEM model is derived excluding the thin catalyst layers and very small fillets due to their negligible contributions to the mechanical properties. The gas connectors and the stoppers are also not regarded in the FEM model as they don't influence the mechanical characteristics of the fuel cell stack.

One eighth of the 3D model is taken into account by using the advantage of symmetry conditions, which results in a proper utilization of the computational capacity. The stack with 5 cells and the derived FEM model are illustrated in Figure 4.7. A single fuel cell in the FEM model consists of the components given in Figure 4.7. As it can be extracted from the Figure 2.5 and Figure 4.7, the finite element model for the fuel cell stack is derived in accordance with the original models. The geometry of the silicone sealing elements are adapted to the thickness of the GDL to get the contact between the components as it occurs in the mounted fuel cell stack. Disc springs are removed from the FEM model assuming the direct transformation of axial loads on washers.

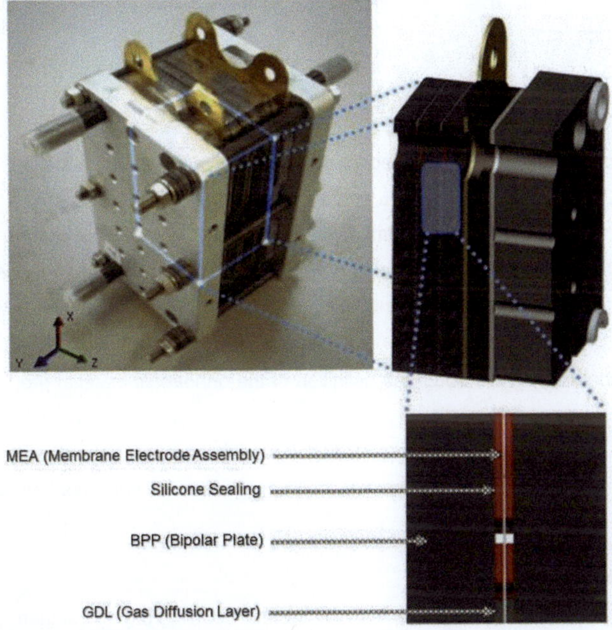

Figure 4.7 Stack with 5 cells and derived FEM model

4.2.2.2 Definition of Physics

The mathematical definition of the physics is derived from the problem statement and corresponds to the boundary settings given in section 4.2.2.3. The mathematical modeling of a physical phenomenon with the help of differential equations follows the implementation of the finite element approach to be able to get a solution as explained in section 4.1.

4.2.2.2.1 Structural Modeling

The governing equation of the structural mechanics is shown in Eq. 4.19. The differential equation defining the structural mechanics is derived from the equation of motion with regard to the equilibrium condition [85, 124].

$$\nabla \cdot \left(D \nabla \underline{u} \right) = 0 \qquad \text{Eq. 4.19}$$

$\underline{u}$, D in Eq. 4.19 represent the deformation vector and stiffness matrix for isotropic materials respectively [125]. The stiffness matrix D is given in Eq. 4.20. E denotes the Young's modulus and v is Possion's ratio.

$$D = \frac{E}{(1+v)(1-2v)} \begin{bmatrix} 1-v & v & v & 0 & 0 & 0 \\ v & 1-v & v & 0 & 0 & 0 \\ 0 & 0 & 0 & \frac{1-2v}{2} & 0 & 0 \\ 0 & 0 & 0 & 0 & \frac{1-2v}{2} & 0 \\ 0 & 0 & 0 & 0 & 0 & \frac{1-2v}{2} \end{bmatrix}$$

Eq. 4.20

The isotropic material modeling is used for the whole components in the fuel cell stack model except the GDL components. A GDL has anisotropic material properties due to its porous structure [126]. The through plane material properties of GDL are different from the in plane directions. Orthotropic material properties are used for the GDL components in the FEM model considering the anisotropy in the through plane direction. The Young's Modulus of the GDL in through plane direction is defined as a function of strain (ε_y) and can be found in section 4.2.2.4. Further information about the definition of the orthotropic material properties can be found in [125].

The accumulated strain in the computational model is calculated with the use of the deformation vector $\underline{u}$ and consists the elastic, thermal and initial strain contributions as given in Eq. 4.21.

$$\varepsilon = \varepsilon_{el} + \varepsilon_{th} + \varepsilon_0$$

Eq. 4.21

The term of thermal strain ε_{th} comes from the thermal coupling due to the thermal expansion and is described in the next subsection.

4.2.2.2.2 Thermal Modeling

In order to get the temperature profile in the ZBT fuel cell stack at stationary cell operation conditions, the thermal modeling is utilized. The governing equations for the heat transfer are given in Eq. 4.22 and Eq. 4.23 with and without convection respectively.

$$\nabla \cdot (-k\nabla T) = Q$$

Eq. 4.22

$$\nabla \cdot (-k\nabla T) = Q - \rho C_p \underline{u} \cdot \nabla T$$

Eq. 4.23

The calculated temperature T builds the temperature profile for the computational domain, which is implemented into the structural module via thermal expansion given in Eq. 4.24.

$$\varepsilon_{th} = \alpha(T - T_{ref})$$

Eq. 4.24

The term α is the thermal expansion coefficient. The calculated temperature T in Eq. 4.22 and Eq. 4.23 is implemented into the structural mechanics module with the reference ambient temperature T_{ref} of 23°C as given in Eq. 4.24. The calculated thermal strain ε_{th} is implemented in the structural mechanics module via Eq. 4.21.

4.2.2.3 Boundary Settings

The computational model developed with the help of CAD platforms is imported into the COMSOL Multiphysics® to build the FEM Model. The corresponding boundary settings are to be defined after definition of the physics acting on the computational domain as explained before. The boundary settings for the structural and thermal modeling are defined in this section separately.

4.2.2.3.1 Structural Modeling

The FEM model consists of one eighth of the original CAD model utilizing the symmetry conditions as explained before. The corresponding symmetry conditions are to be defined in the structural mode for the reduced geometry. The highlighted boundaries illustrated in the Figure 4.8-(a) are the symmetry conditions for the FEM model. The cutting surfaces to generate the FEM geometry are the symmetry conditions as it can be seen in Figure 4.8.

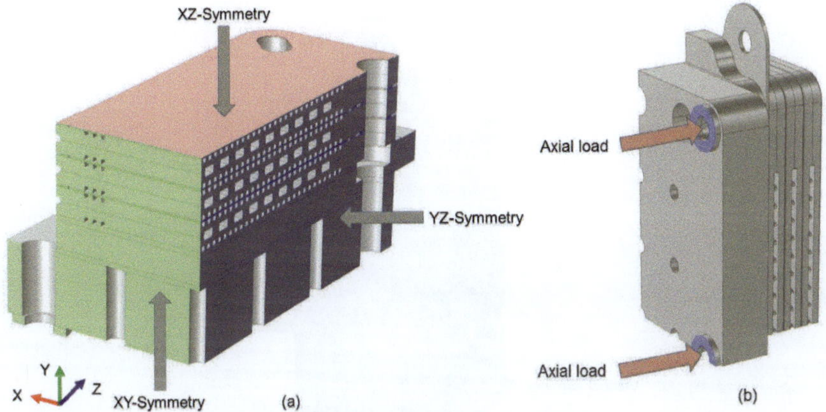

Figure 4.8 Symmetry conditions (a) and boundary conditions (b) on the FEM model

The axial load acting on the washers due to the tie rods is calculated as 1250[N] according to [127, 128]. The calculated axial load is assigned to the highlighted boundaries on the washers as depicted in Figure 4.8-(b).

The grub screws considered in the FEM model are not regarded in the simulations made in this section. The grub screws are not commonly used in standard fuel cell applications. The results of the simulations considering the grub screws are presented in [119].

4.2.2.3.2 Thermal Modeling

Thermal modeling of the fuel cell stack is performed with the boundary settings given in this subsection. For the thermal constraints the fuel cell stack operation temperature is utilized. The stationary operation of the fuel cell stack with a constant temperature value is assumed neglecting the reaction dependency through membrane. A constant temperature value of 70 [°C] is assigned to the membranes in the FEM model.

The convective constraints acting on the fuel cell stack are modelled with the help of the heat transfer library of COMSOL Multiphysics® [129].The governing boundary equation for the convective heat transfer is given in Eq. 4.25.

$$-\boldsymbol{n} \cdot (-k\nabla T) = h\left(T_{inf} - T\right)$$

Eq. 4.25

T is the calculated temperature value on the model. T_{inf} is the ambient temperature and has a value of 23°C. The term h is the heat transfer coefficient depending on boundary definition and realized with the help of empirical approximations. n is the normal vector and represents the boundary normal.

The air cooling with the help of a radial fan is modelled with the help of forced convection constraint in the FEM model. The boundaries, on which the forced convection is acting, are shown in Figure 4.9 highlighted in green.

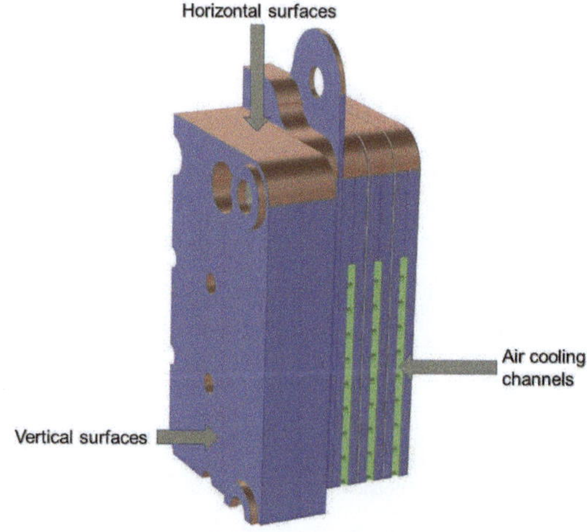

Figure 4.9 Free convection and convective constraints on the FEM model

The heat transfer coefficient term h for the forced convection in enclosed isothermal channels is given in the Eq. 4.26 [129].

$$h = \begin{cases} \dfrac{k}{D}3.66 & Re_D \leq 2500 \\[2ex] \dfrac{k}{D}0.027 Re_D^{4/5} Pr^{0.4} \left(\dfrac{\mu}{\mu(T)}\right)^{0.14} & Re_D > 2500 \end{cases}$$

Eq. 4.26

D is the tube diameter. The heat transfer coefficient for the forced convection is calculated with the help of cooling air velocity, which is predefined according to the used radial fan properties. The free convection acting on the fuel cell stack at the outer surfaces is also modelled with the help of the COMSOL Multiphysics® heat transfer library. The selected boundaries, on which the horizontal and vertical free convection constraints are acting, are depicted in Figure 4.9 and highlighted in red and blue respectively.

The mathematical definition of the vertical free convection is given in Eq. 4.27.

$$
h = \begin{bmatrix}
\dfrac{k}{L}\left(0.68 - \dfrac{0.67 Ra_L^{1/4}}{\left(1 + \left(\dfrac{0.492k}{\mu C_p}\right)^{9/16}\right)^{4/9}}\right) & Ra_L \leq 10^9 \\[4em]
\dfrac{k}{L}\left(0.825 + \dfrac{0.387 Ra_L^{1/6}}{\left(1 + \left(\dfrac{0.492k}{\mu C_p}\right)^{9/16}\right)^{8/27}}\right) & Ra_L > 10^9
\end{bmatrix}
\qquad \text{Eq. 4.27}
$$

The mathematical definition of the horizontal free convection is given in Eq. 4.28.

$$
h = \begin{bmatrix}
\left(\dfrac{k}{L}0.54 Ra_L^{1/4}\right) & Ra_L \leq 10^7 \\[1.5em]
\left(\dfrac{k}{L}0.15 Ra_L^{1/3}\right) & Ra_L > 10^7
\end{bmatrix}
\qquad \text{Eq. 4.28}
$$

Further information about the mathematical definition of the heat transfer coefficient with the help of COMSOL Multiphysics® can be found in [129, 130].

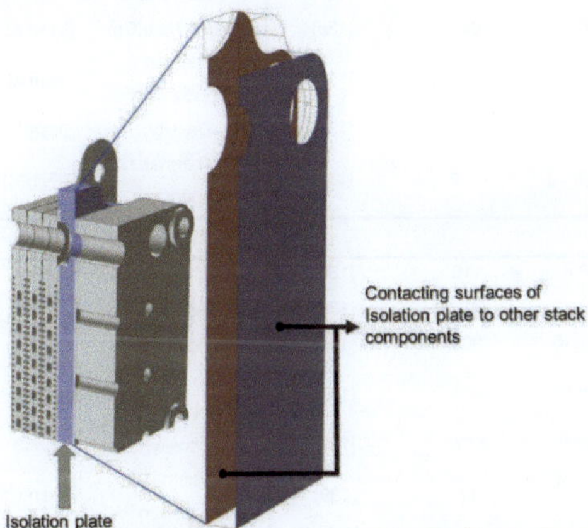

Figure 4.10 Thermal contact resistance constraints

Thermal contact resistance is also considered on the contacting surfaces of the isolation plate as depicted in Figure 4.10. The mathematical definition of the thermal contact resistance is given in Eq. 4.29.

$$
-n_d \cdot (-k_d \nabla T_d) = -\frac{T_u - T_d}{R_s}
\qquad \text{Eq. 4.29}
$$

The value of thermal contact resistance R_s is defined as 0.02 [K.m²/W] according to [131-133]. The subscripts u and d represent the upside and downside of contacting surfaces respectively.

4.2.2.4 Material Properties

The fuel cell stack components and their properties are handled in detail in section 2.3.2. The component list of the FEM model and the assigned material properties can be found in Table 4.1.

Table 4.1 Component and material list

Material Properties	Unit	Value	Source	Material Properties	Unit	Value	Source
Alu-5083-Endplate				**Nafion-N-112**-Membrane			
Young's Modulus	[GPa]	71	[134]	Young's Modulus	[GPa]	0.249	[65]
Poisson's ratio		0.33	[134]	Poisson's ratio		0.25	[137]
Specific Heat(Cp)	[J/kg.K]	899	[134]	Specific Heat(Cp)	[kJ/kg.K]	4188	[138]
Thermal Conductivity	[W/m.K]	117	[134]	Thermal Conductivity	[W/m.K]	0.18	[138]
Density	[kg/m³]	2660	[134]	Density	[kg/m³]	1970	[65]
Thermal Expansion	10^{-6} [1/°C]	23.8	[134]	Thermal Expansion	10^{-6} [1/°C]	123	[137]
Graphite compound(77.9%-Graphite)-Bipolar plate				**POM**-Isolation plate			
Young's Modulus	[GPa]	15.56	[135, 136]	Young's Modulus	[GPa]	3.1	[142]
Poisson's ratio		0.25	[137]	Poisson's ratio		0.35	[134]
Specific Heat(Cp)	[kJ/kg.K]	685	[138]	Specific Heat(Cp)	[kJ/kg.K]	1.5	[134]
Thermal Conductivity	[W/m.K]	25,15,25	[135]	Thermal Conductivity	[W/m.K]	0.31	[134]
Density	[kg/m³]	1780	[135]	Density	[kg/m³]	1410	[142]
Thermal Expansion	10^{-6} [1/°C]	12.2	[135]	Thermal Expansion	10^{-6} [1/°C]	110	[142]
Copper-Current collector				**Silicone(Wacker-RT 624)**-Sealing			
Young's Modulus	[GPa]	110	[139]	Young's Modulus	[MPa]	$f_s(\varepsilon_y)$	[135]
Poisson's ratio		0.35	[139]	Poisson's ratio		0.49	[140]
Specific Heat(Cp)	[kJ/kg.K]	385	[139]	Specific Heat(Cp)	[kJ/kg.K]	0.2	[144]
Thermal Conductivity	[W/m.K]	400	[139]	Thermal Conductivity	[W/m.K]	1175	[143]
Density	[kg/m³]	8700	[139]	Density	[kg/m³]	1070	[144]
Thermal Expansion	10^{-6} [1/°C]	17	[139]	Thermal Expansion	10^{-6} [1/°C]	170	[144]
Dry GDL(Toray paper)-Gas diffusion layer				**Steel 1.4307-ASI 304**-Pressure plate, Flat washer			
Young's Modulus	[GPa]	10, $f_{GDL}(\varepsilon_y)$, 10	[126,135, 137]	Young's Modulus	[GPa]	193	[141]
Poisson's ratio		0.25, 0, 0.25	[126, 137]	Poisson's ratio		0.3	[141]
Specific Heat(Cp)	[kJ/kg.K]	685	[138]	Specific Heat(Cp)	[kJ/kg.K]	500	[141]
Thermal Conductivity	[W/m.K]	21,1.7,21	[145]	Thermal Conductivity	[W/m.K]	15	[141]
Density	[kg/m³]	2045	[138]	Density	[kg/m³]	8000	[141]
Thermal Expansion	10^{-6} [1/°C]	-0.8	[145]	Thermal Expansion	10^{-6} [1/°C]	16	[141]

Stack components exhibit isotropic material properties excluding the bipolar plates and gas diffusion layers. The orthotropic material properties of the GDL in the FEM model are discussed in section 4.2.2.2.1. Compression stiffness of the GDL increases as a result of the porous structure. In order to define the material properties of the GDL, the thickness of a dry GDL (Toray Carbon Paper) has been measured with an [g]INSTRON® testing Instrument. The thickness of GDL under varying compression pressure is given in Figure 4.11.

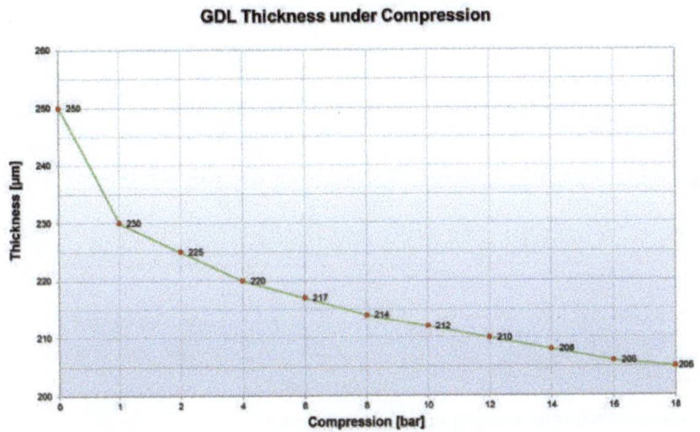

Figure 4.11 GDL thickness under compression

The modulus of elasticity of the GDL is calculated with the help of the measured data. The corresponding Young's Modulus is fitted as a function of the through plane strain with the help of MATLAB® script and given in Eq. 4.30.

$$f_{GDL}(\varepsilon_y) = (0.49e^9\varepsilon_y^2) - (0.041e^9\varepsilon_y) \qquad \text{Eq. 4.30}$$

The calculated modulus of elasticity and fitted function $f_{GDL}(\varepsilon_y)$ are depicted in Figure 4.12.

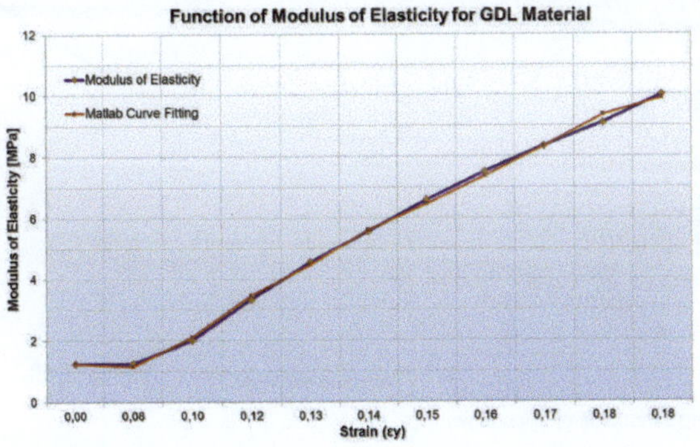

Figure 4.12 Young's Modulus of GDL

---

[g]Trademark of Illinois Tool Works Inc. (ITW)

The elasticity function $f_{GDL}(\varepsilon_y)$ of the GDL is implemented as a subdomain variable and assigned to the GDL orthotropic material properties.

Mechanical properties of the silicone sealing is defined in the same manner like GDL components by measuring the thickness of the silicon sealing under compression. The thickness of silicone sealing under varying compression pressure is given in Figure 4.13.

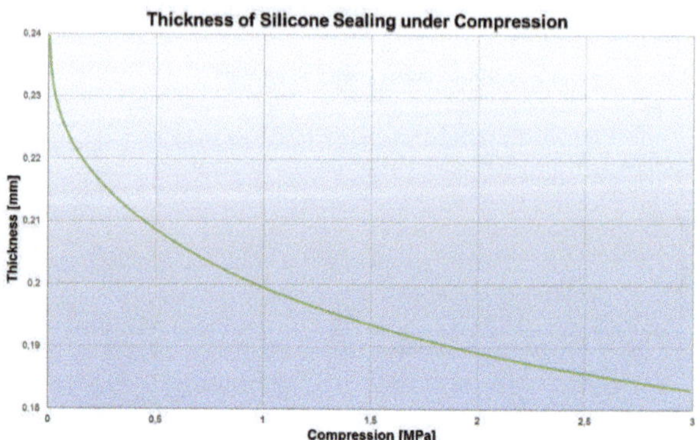

Figure 4.13 Thickness of silicone sealing under compression

The modulus of elasticity of silicone is generated with the help of the measured data. The fitting function for the Young's Modulus of silicone is given in Eq. 4.31.

$$f_s(\varepsilon_y) = \left(434.79e^6\varepsilon_y^2\right) + \left(42.95e^6\varepsilon_y\right) + 5.93e^6 \qquad \text{Eq. 4.31}$$

The calculated modulus of elasticity and fitted function $f_s(\varepsilon_y)$ of the silicone sealing are depicted in Figure 4.14.

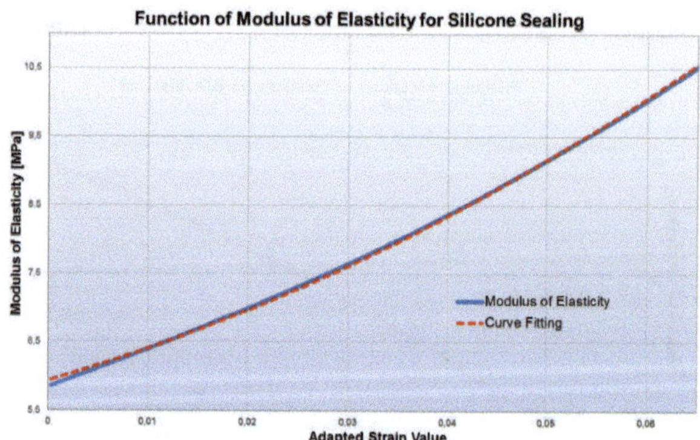

Figure 4.14 Young's Modulus of silicone sealing

The elasticity function $f_s(\varepsilon_y)$ of silicone sealing is assigned as a subdomain variable for corresponding components analogous to the GDL components.

4.2.2.5 Meshing and Solver Settings

Meshing of the FEM model is to be made for the computations as explained in section 4.1. Several mesh configurations are performed to ensure the independency of the results from the meshing properties. Mapped, swept and free meshing options are used depending on the geometry, stress configuration and state of concern.

The final mesh configuration taken into account by the performed simulations can be seen in Figure 4.15. The crucial components to be analyzed in detail are meshed finer taking the computational capacity into consideration. The compression of the cell components is the point of interest. Thus the meshing of cell components is built precisely and given in zoom box in Figure 4.15 for a detail view. This leads to finer results due to the geometrical and material properties.

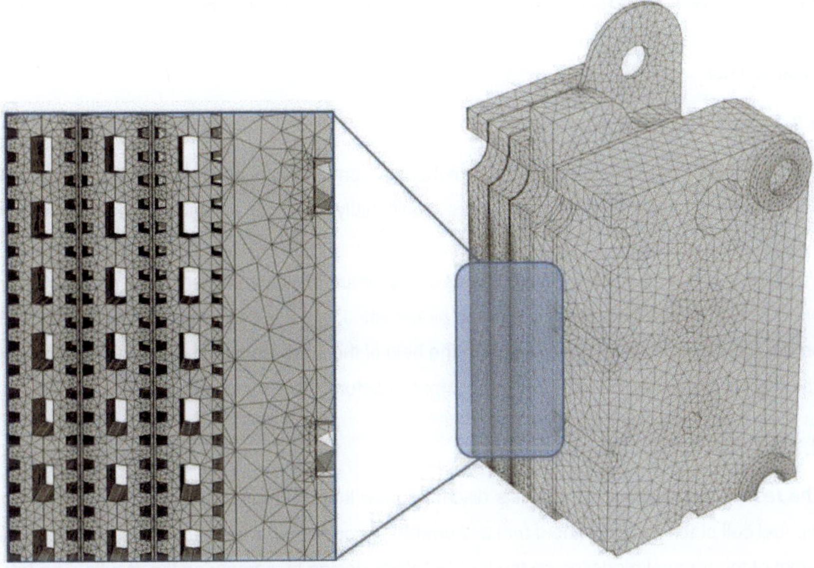

Figure 4.15 Detailed view of meshing elements

Meshing properties for both structural and thermomechanical computations can be seen in Table 4.2.

Table 4.2 Meshing properties

	Number of Elements	Degree of Freedom	Element Quality
Structural	557,335	4,729,272	0.4674
Thermomechanical	557,335	6,310,283	0.4674

The meshing properties are kept the same for both structural and thermomechanical computations.

The linear and quadratic type of elements are selected for the FEM model due the computational capabilities and modeling properties. Using quadratic type of elements increases the computation time and memory requirements. On the other hand utilizing quadratic elements approximates the results finer. Quadratic type of element is assigned for the structural computations due to the existing nonlinearity in the modeling and the point of interest. The linear type of elements is assigned to the thermal calculations, which can be noticed by the number of degree of freedom in Table 4.2.

With regard to the complexity of the 3D model and the amount of the degree of freedom, an iterative solver (GMRES) is chosen. For the thermomechanical computations including both thermal and structural analysis, a segregated solver is selected. In segregated solver, the iterative solver (GMRES) is assigned to both modules for the computation of the variables. Further information about meshing and solver settings with COMSOL Multiphysics® can be found in [147].

4.2.3 Results and Evaluation

The results of the structural and thermomechanical computations are presented in this section. The results of the thermal solutions are additionally given in thermomechanical analysis subsection.

Von Mises stress and displacement values of the simulations are presented for the evaluation of the results. Von Mises stress is a scalar stress value, which represents the equivalent stress on the components and is computed with the help of the stress tensor [146]. The displacement value is generated with the help of the computed deformation vector as given in Eq. 4.19.

4.2.3.1 Structural Analysis

The results of the structural analysis discarding the thermal modeling enable the analysis of the fuel cell stack design without fuel cell operating conditions. It also ensures to evaluate the effect of the thermal modeling on the fuel cell stack design by comparing the results. The total displacement and compression of the fuel cell stack components can be seen in Figure 4.16. The total displacement value is scaled in Figure 4.16-(a) to be able to realize the deformation on the fuel cell stack. The wireframe of the original geometry is also visualized to compare with the fuel cell stack deformation.

The deformation on the endplate is given in Figure 4.16-(b) separately. The deflection intensifies on the corners of the endplate due to the position of the acting loads and boundary conditions as expected. The value of the total displacement on the corners of the endplate reaches to a value of 0.0938[mm].

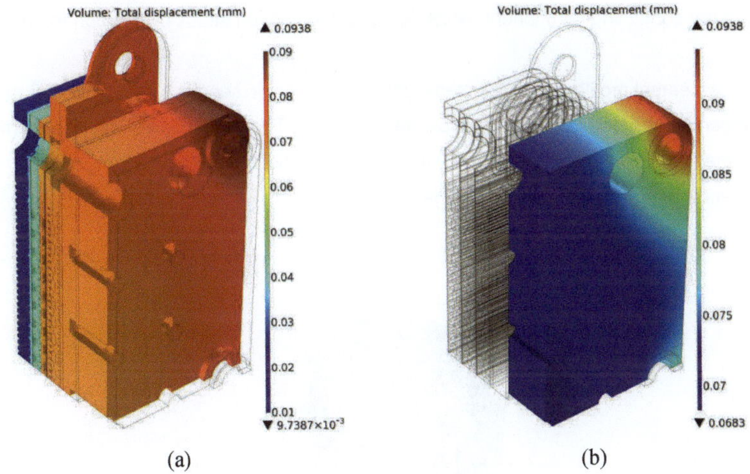

Figure 4.16 Total displacement [mm] for structural analysis (a) scaled on the fuel cell stack (b) on the endplate

In order to build the whole fuel cell stack geometry, the results are mirrored on the symmetry boundaries and given in Figure 4.17 with a scaled size. Hence the deflection of the whole fuel cell stack geometry can be seen and analyzed in detail. The deformation on the corners and in the middle of the endplate can be captured properly.

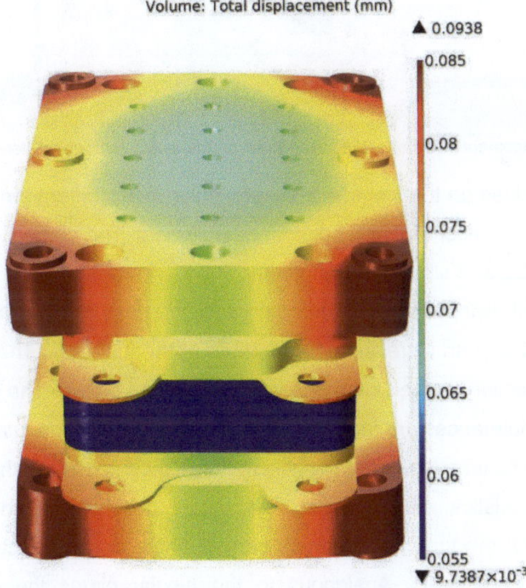

Figure 4.17 Scaled total displacement [mm] for structural analysis on the whole fuel cell stack

The deflection difference on the endplate contributes to the buckling of the endplate. A strong buckling of the endplates results in the transformation of the compression forces through the edges of the contacting component, which leads to the uneven stress distribution for the fuel cells.

Total displacement profile on the selected bipolar plates can be seen in Figure 4.18. The range of the legend is adapted to be able to evaluate the results. The distribution of the displacement field on the first bipolar plate is formed by the load distribution acting on the fuel cell stack and the contact with the neighboring geometries.

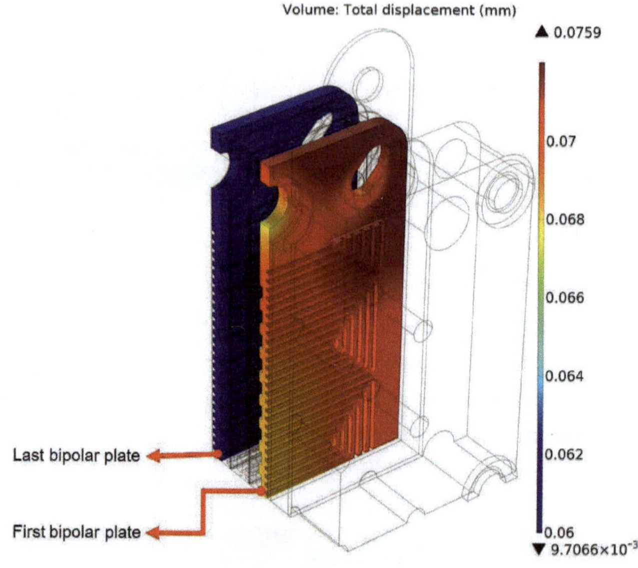

Figure 4.18 Total displacement [mm] for structural analysis on the selected bipolar plates

The displacement values on the first and the last bipolar plates differ from each other due to the total cumulative displacement and boundary constraints. The components between both bipolar plates like GDL and silicone sealing components contribute to the homogenization of the total displacement distribution. A better homogenized stress distribution in the middle fuel cell can be expected with an increase in the number of the fuel cells. The bipolar plates are compound based injection molded components as explained before. The surface roughness and the geometrical tolerances are formed due to the material and manufacturing properties. The planarity of the bipolar plates are diverging in the real components, that leads to uneven compression characteristics, which should be considered by evaluation of the simulation results. In Figure 4.19 the total displacement can be seen for the selected membranes by suppressing other components. Analogous to the bipolar plates the distribution of the membrane deflection neutralizes in the middle cell. The load transformation to the membranes occurs through the silicone sealing components and GDL, which can be realized in Figure 4.19.

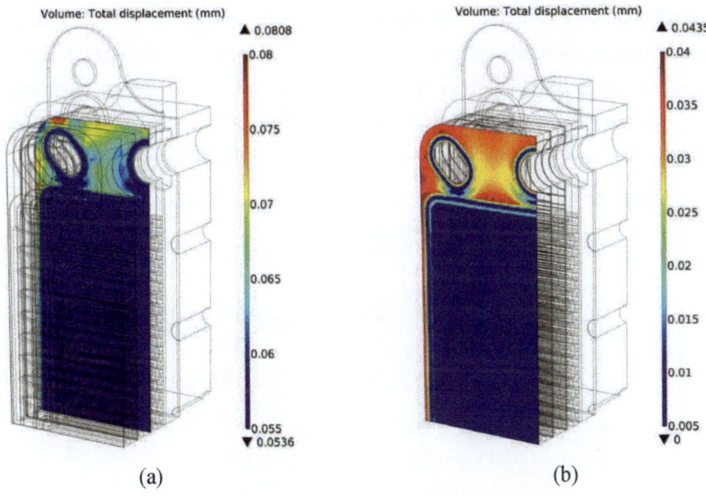

(a)   (b)

Figure 4.19 Total displacement [mm] for structural analysis on the selected membranes

Von Mises stress can be seen in Figure 4.20 for the fuel cell stack and the bipolar plates respectively. As expected there is higher stress resolution in the regions, where the loads are acting and contact between components occurs. Von Mises stress values on components don't exceed the critical stress values, which can cause cracks and undesired material fractures. This is also not observed in the real fuel cell stack.

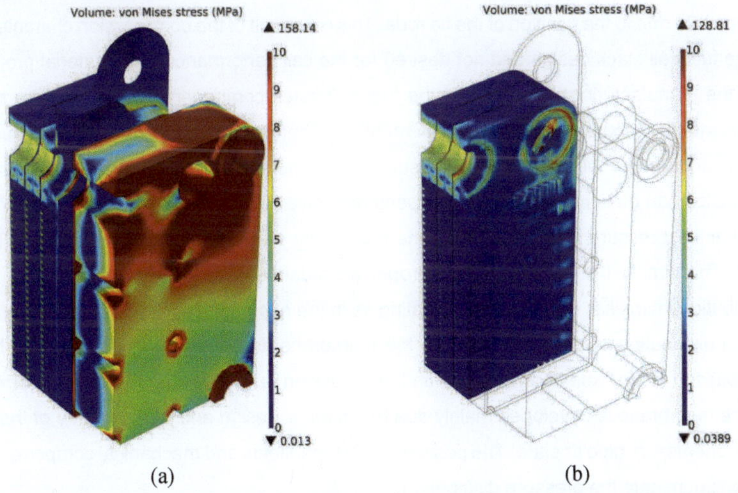

(a)   (b)

Figure 4.20 Von Mises stress [MPa] for structural analysis (a) on the fuel cell stack (b) on the bipolar plates

The stress distribution in Figure 4.20 provides an additional tool to analyze and conceive the sealing design due to the determination of the higher stress regions. In Figure 4.21 Von Mises stress distribution on a selected membrane is given with an adapted legend. The upper side of the membrane is zoomed and given in Figure 4.21 together with a revised legend, in order to analyze the stress distribution on the membrane caused by the contact with the sealings. The stress distribution on the selected membrane intensifies on the border of the silicone sealing components due to the load transformation through the sealing components.

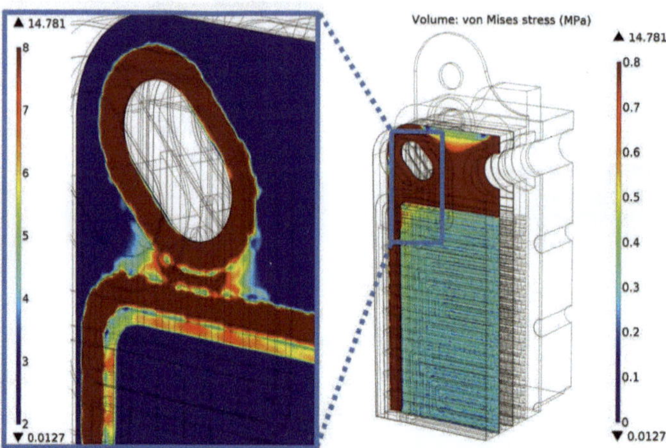

Figure 4.21 Von Mises stress [MPa] for structural analysis on the selected membrane

The upper side of the active area on the membrane is under more compression as seen in Figure 4.21 due to the position of the tie rods. This is a result of the compression characteristics of the fuel cell stack design and not desired for the cell performance. The material properties and the manufacturing tolerances of the fuel cell stack components particularly the bipolar plates modify the stress configuration as mentioned before.

The distribution of the mechanical pressure on the membrane without the effect of thermal expansion can be compared with the experiments given in section 3.2. The distribution of the mechanical pressure is visualized with the help of the pressure measurement films in Figure 3.13. The simulation results are in a proper accordance with the experiments. The stress distribution intensifies on the silicone sealing as in the experimental results. The affinity of the stress distribution on the upper side of the membrane active region can be noticed by the comparison of the experimental data with the simulation results. Mechanical stress distribution on the membrane is developed mainly due to the stack design and the geometry of the stack components e.g. bipolar plate. The positions of the manifolds and mechanical components e.g. tie rods dominate the pressure distribution.

In order to analyze the stress configuration and the compression characteristics of the fuel cell stack design, the GDL components are investigated in detail. As mentioned before, the GDL and GDL compression characteristics play an important role in fuel cell performance

[120, 122]. In Figure 4.22 the high scaled through plane displacement of selected GDLs of the first and the last fuel cell is figured out for the structural analysis. The bipolar plate is also given in Figure 4.22 together with the simulation results in order to realize the flow field structure and position of the GDL component properly. The squeezing of the GDL into the flow field channels and the flow field structure can be noticed in Figure 4.22. The squeezing of the GDL into flow fields differs from channel to channel. The value of the through plane displacement and the squeezing of the GDL change depending on the cell position and the Z-direction.

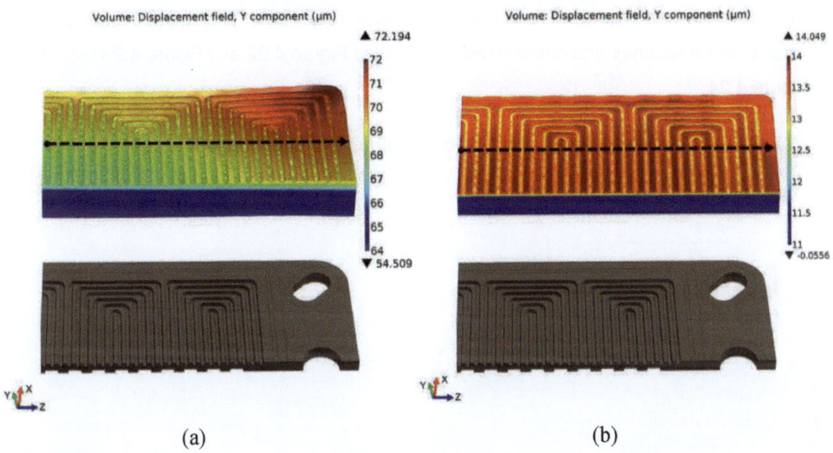

Figure 4.22 Scaled through plane displacement [mm] for structural analysis on the selected GDLs of the (a) first and (b) third fuel cell

The uneven deformation on GDL of the first cell along the Z-axis depends on the position of the acting compression load. Comparing with the other results presented in this section, the squeezing of the GDL in the third fuel cell is also homogenized similar to the other stack components.

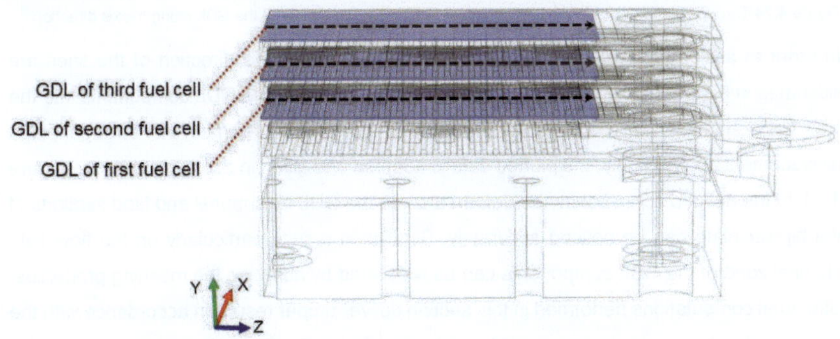

Figure 4.23 Selected GDL components and the defined lines in Z-axis direction

The value and the distribution of the squeezing of the GDL components are in an accordance with the geometrical and material properties. In order to analyze the squeezing phenomenon properly, one GDL component of each fuel cell in the stack is selected as illustrated in the Figure 4.23.

Three different lines are defined, which begin from the middle of the lowest edges of the corresponding GDLs and lie along the Z-axis direction as seen in the Figure 4.23. The defined line is also illustrated in Figure 4.22 in black directly on the GDL components, in order to assist the understanding of Figure 4.23. The through plane displacements of the GDL components through the defined lines and direction as displayed in Figure 4.22 and Figure 4.23 are plotted in Figure 4.24.

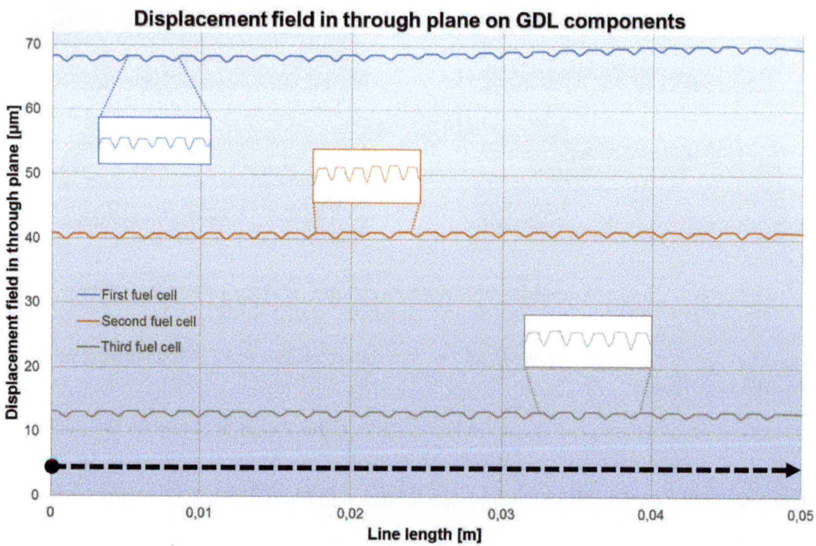

Figure 4.24 Through plane displacement for structural analysis in the middle of the GDL along Y-axis direction

In order to assist the evaluation of Figure 4.24, the position and direction of the lines are illustrated in Figure 4.24 with the black arrow. The squeezing of the GDL components into the flow field channels can be realized properly. In order to investigate the through plane displacement appropriately, the plotted data is adapted and given in zoom boxes in the Figure 4.24. In the defined zoom boxes the deformation on the GDL by channel and land sections of the bipolar plate can be noticed accurately. Simulation results particularly on the flow field channel zone of the GDL components can be recovered by adapting the meshing properties. Structural computations performed in this section deliver proper results in accordance with the experiments and provide an appropriate understanding of the fuel cell stack clamping [148-150].

4.2.3.2 Thermomechanical analysis

The results of the thermomechanical computations are presented in this subsection. The modeling and the boundary settings for the calculations are defined in sections 4.2.2.2 and 4.2.2.3 respectively. The difference from the results presented in previous section is the consideration of the thermal expansion on the fuel cell stack. The results of the thermomechanical computations enable the analysis of the fuel cell stack design within the fuel cell operating conditions.

The results of the performed thermomechanical computations can be compared with the experimental results presented in section 3 for the verification purposes. The results in this section are also to be compared with the results in previous section to analyze the effect of the thermal expansion on the mechanical characteristics of the fuel cell stack.

The figures presented in this section are given with the same properties and sequence like in previous section to simplify the comparison and analysis of the results. In Figure 4.25-(a) the temperature profile of the fuel cell stack is figured out.

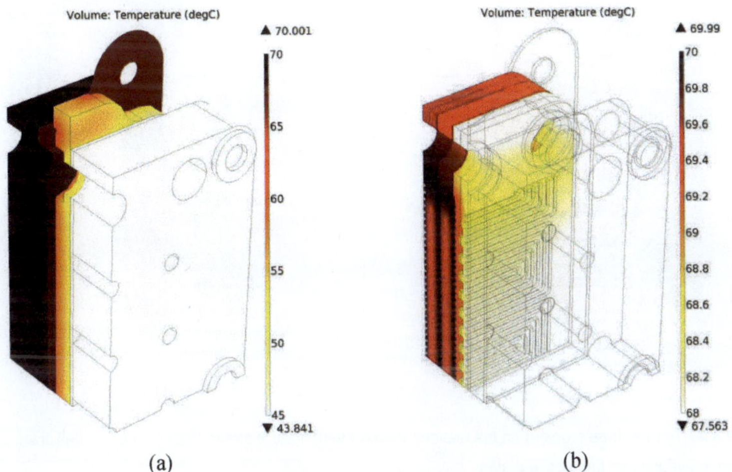

Figure 4.25 Temperature profile [°C] (a) of the fuel cell stack (b) of the bipolar plates

The temperature gradient changes through the isolation plate as expected. Temperature values in the endplate occur due to the heat transfer through the isolation plate and cooling via free convection.

Thermal distribution in the bipolar plates is displayed in Figure 4.25-(b) by suppressing other stack components and adapting the legend. Thermal distribution in the bipolar plates is developed due to the heat transfer through the neighboring components and cooling channels. The difference in the temperature values of the first and third cell occurs due to the heat transfer through the isolation plate. The temperature gradient in the cooling channels and membrane zones of the bipolar plates can be captured.

The stream lines and isothermal surfaces are utilized as additional post processing options in the Figure 4.26 to investigate the change in temperature of the fuel cell stack. In the Figure 4.26-(a) the temperature flow on the fuel cell stack can be seen by following the ribbons. The temperature loss via the outer surfaces of the stack can be realized. The temperature loss on the bipolar plates via the cooling channels can be noticed In Figure 4.26-(b) with the help of the isothermal surfaces.

The fuel cell stack temperature profile derived from the performed simulations are in a proper accordance with the measured temperature values of the single points on the fuel cell stack, which are presented in section 3. The temperature values of the endplate and bipolar plates achieved by the thermal measurements assure the temperature distributions in the thermal simulations.

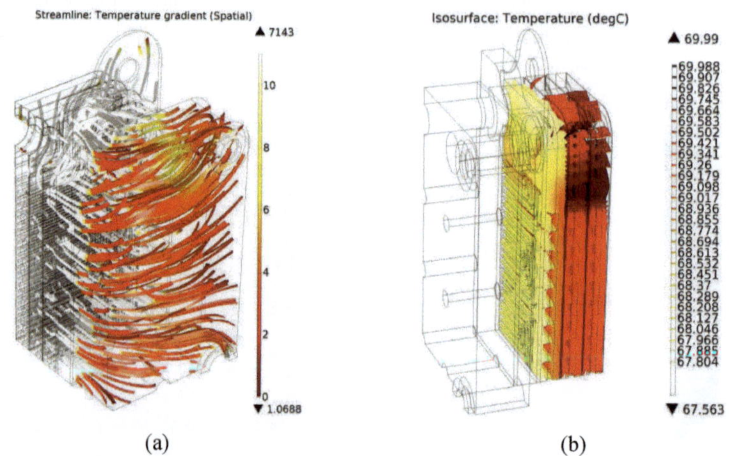

Figure 4.26 Temperature gradient on the fuel cell stack(a) temperature streamlines on the fuel cell stack (b) isothermal surfaces on the bipolar plates

In the real case the temperature distribution in the membrane depends on the reaction rate occurring on the membrane surfaces. The regulation of the cooling fan and the flow distribution of the cooling air are also assumed to be constant by the performed computations.

Including the thermal expansion, the scaled total displacement profile of the fuel cell stack is given in Figure 4.27. The thermal expansion of the fuel cell stack can be realized by comparing the distribution profile of the total displacement with Figure 4.16. The stack expands due to the thermal expansion of the fuel cell stack components. The thermal expansion of the stack components differ from each other according to their geometries, thermal expansion coefficients and temperature profiles.

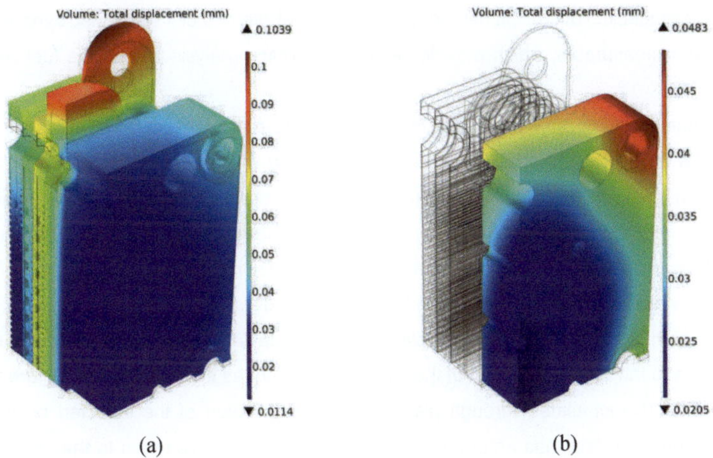

Figure 4.27 Total displacement [mm] for thermomechanical analysis (a) scaled on the stack (b) on the endplate

The deformation on the endplate is given in Figure 4.27-(b) separately. The change in the deflection distribution of the endplate can be noticed by comparing with the Figure 4.16-(b). The outwards elongation of the endplate due to the thermal expansion can be realized.
Total displacement of the whole fuel cell stack is built in Figure 4.28 analogous to the Figure 4.17.

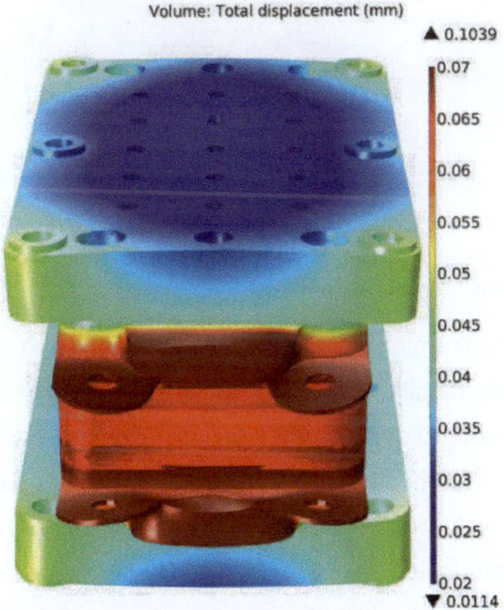

Figure 4.28 Total displacement [mm] for thermomechanical analysis on the whole fuel cell stack

The deflection intensifies on the corners of the endplate similar to the Figure 4.17. However the thermal expansion changes the displacement distribution and balances the deformation on the fuel cell stack. Total displacement profile of the whole fuel cell stack displayed in Figure 4.28 is not measurable, which provides a supplementary analysis tool for the fuel cell stack design.

The simulated displacement value in the middle of the outer surface of the endplate can be verified with the measured data presented in section 3. In order to analyze the thermal expansion properly and verify the simulations, the difference between the structural and thermomechanical results is derived and depicted in Figure 4.29. In other words Figure 4.29 is the difference between Figure 4.28 and Figure 4.17 derived for the through plane (Y-axis) direction, which was measured with the help of a dial extensometer in section 3.

A single point in the middle of the endplate outer surface is selected as shown in Figure 4.29 analogous to the measuring point of the dial extensometer as it can be seen in Figure 3.1 and Figure 3.2. The simulated through plane displacement value of the selected point on the endplate outer surface has an approximate value of 48[μm]. This refers to the displacement for one side of the fuel cell stack.

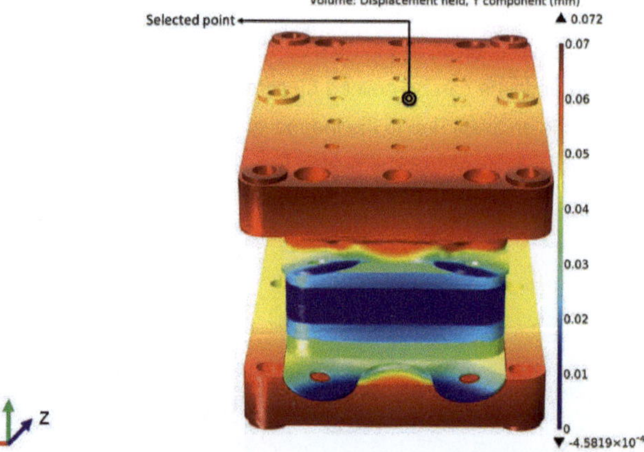

Figure 4.29 Displacement field in through plane(Y-axis) [mm] for thermal expansion of the whole fuel cell stack

The FEM model is free to thermally expand in both directions in Y-axis due to the boundary conditions. The simulation results deliver the displacement values on each single endplate. Hence the experimental set-up delivers the total displacement value for the whole stack as explained in section 3.1. The measured thermal deflection of the whole fuel cell stack has a cumulative approximate value of 94[μm] as plotted in Figure 3.3. Considering both of the endplates together, the simulation results for the whole fuel cell stack deflection contributes to an approximate value of 96[μm], which shows a proper accordance with the measured data. Thermal modeling, material properties, geometrical tolerances and measurement errors are potential reasons for this negligible difference of about 2.1% between the simulation results and measured data.

Two bipolar plates are selected and their total displacement values are displayed in Figure 4.30 by suppressing other stack components.

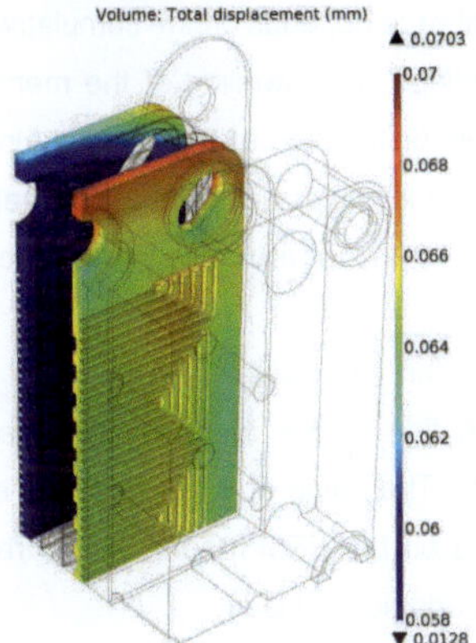

Figure 4.30 Total displacement [mm] for thermomechanical analysis on the selected bipolar plates

The wireframe of the stack geometry is given to remark the positioning of the bipolar plates. The distribution of the total deflection in the bipolar plates is changed comparing with Figure 4.18. This is caused by the thermal expansion of the bipolar plates, which depends on the temperature distribution and geometry of the plates and contributes to an even distributed displacement via accumulated strain. Total displacement profile on the selected membranes can be seen in Figure 4.31 by suppressing other components.

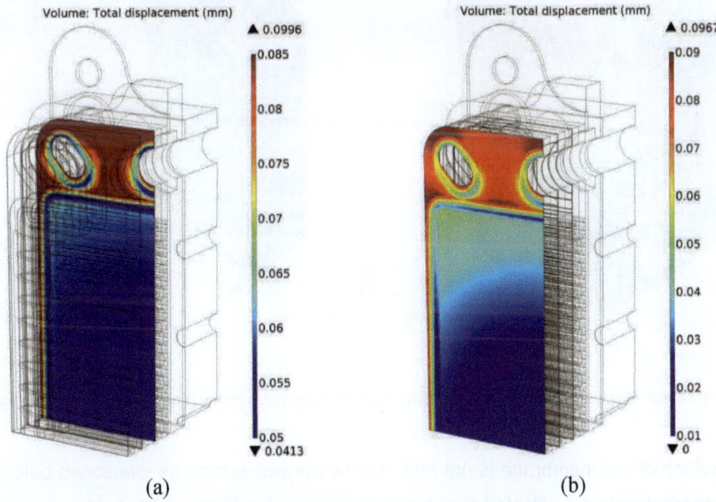

(a)   (b)

Figure 4.31 Total displacement [mm] for thermomechanical analysis on the selected membranes

Thermal expansion of the components changes the distribution of the total displacement in the membranes comparing with Figure 4.19. The difference in the total displacement distribution between the first and third membrane depends on the cumulative thermal expansion of the cell components and boundary settings. The swelling of the membrane is not regarded by the simulations due to the unknown distribution of the humidity in membranes and the relative significance of the swelling on the total displacement of the fuel cell stack comparing with the thermal expansion. The effect of the swelling on the deflection of the fuel cell stack is investigated in section 3.1.1.1.

Von Mises stress distribution of the fuel cell stack and bipolar plates is given in Figure 4.32 respectively. The stress distribution on the fuel cell stack seems to stay almost identical comparing with the Figure 4.20. The value of the Von Mises stress is increased due to the additional thermal stress acting on the FEM model. Stress resolution is intensifying in the regions of the contact surfaces and acting loads.

In Figure 4.32-(b), the stress distribution on the bipolar plates is displayed. The components are forced to expand outwards due to the temperature change and this can be observed for the bipolar plates. The accumulated stress occurring on the bipolar plates is increased as it can be depicted from Figure 4.32-(b).

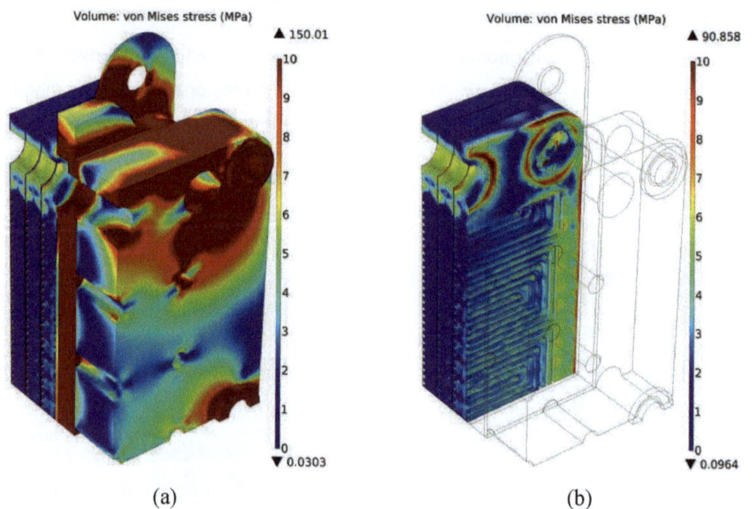

(a)   (b)

Figure 4.32 Von Mises stress [MPa] for thermomechanical analysis (a) on the fuel cell stack (b) on the bipolar plates

The swelling of the membrane is not regarded by the simulations as mentioned before. The stress distribution in the selected membrane is presented in Figure 4.33. Analogous to Figure 4.21.the upper side of the membrane is zoomed and given with a revised legend.

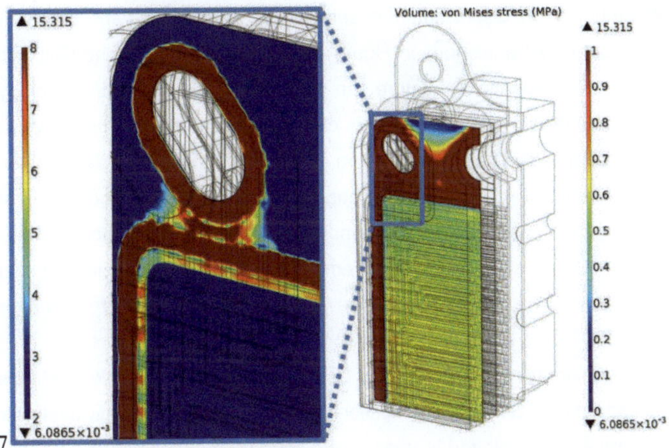

Figure 4.33 Von Mises stress [MPa] for thermomechanical analysis on the selected membrane

As it can be simply noticed by Figure 4.33, the spatial stress distribution on the membrane changes dramatically. Thermal expansions of the membrane and bipolar plates force the contact between components, which results in increase of the stress values on the membrane. It also changes the stress distribution in the middle of the membrane as it can be revealed from Figure 4.33. This contributes to an improved stress distribution on the membrane active area, which is also preferred for the fuel cell performance.

The GDL components are examined in Figure 4.34 for a better understanding of the compressing characteristics of the fuel cell stack design under thermal loading.

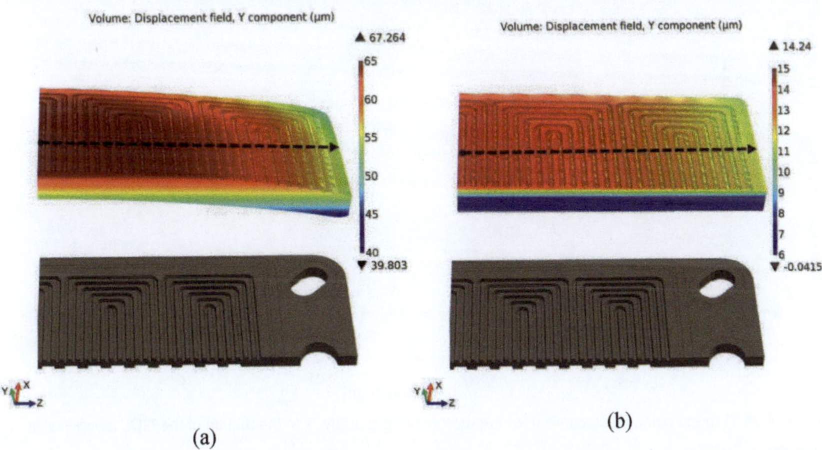

Figure 4.34 Scaled through plane displacement [mm] for thermomechanical analysis on the selected GDLs of the (a) first and (b) third fuel cell

High scaled through plane displacement of selected GDLs of the first and the last fuel cell is figured out in Figure 4.34 analogous to Figure 4.22. The squeezing of the GDL into the flow field channels and the flow field structure can still be captured comparing with Figure 4.22. However the distribution of the squeezing is changed due to the thermal expansion taking place on the components. Thermal expansion of the GDL neighboring components leads to the change in the deflection of the GDL. The negative thermal expansion coefficient of the GDL plays also an important role in the distribution. The thermal expansion occurs from the inner region towards outer surfaces due to the thermal distribution and geometrical shape of the components. This results in an increase of the GDL compression in the middle region, which can be observed in Figure 4.34.

As in previous section the squeezing of the GDL components into the flow field channels is investigated additionally. Simulated deflections on the GDL components along the defined lines as in Figure 4.23 are plotted to analyze the squeezing phenomenon properly in Figure 4.35. Analogous to the Figure 4.24 and Figure 4.34, the squeezing of the GDL through the stacking direction can be realized. The plotted data is adapted and given in zoom boxes in Figure 4.35. In the defined zoom boxes the deformation on the GDL by thermal structural analysis can be noticed properly. BPP lands and flow field channels can still be recognized simply.

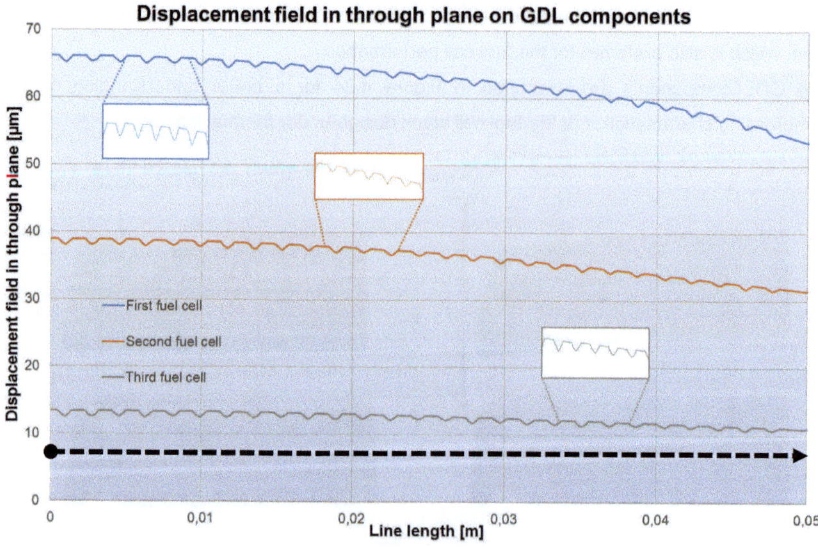

Figure 4.35 Through plane displacement for thermomechanical analysis in the middle of the GDL along Y-axis direction

The results achieved in this section are verified with the help of the measured data given in section 3 by considering the manufacturing tolerances and the precision of the test equipment.

Regarding the test equipment restrictions the results presented in this section show a proper accordance with the measured data performed in section 3. The computational accuracy due to the simulation properties is to be also considered by the evaluation of the results. The thermal simulations can be developed with the help of supplementary measurements. The calculated axial force acting on the washers depends also on several parameters and assumptions, which leads to the imprecision. The definition of the material properties can be improved with the help of additional measurements under varying temperature.

In this section, the presented model enables the analysis of the fuel cell stack mechanics. Thermomechanical analysis of the fuel cell stack with the help of the computational methods provides supplementary tools to study the fuel cell stack design under operating conditions. The deflection and the stress development on the stack components i.e. membrane and gas diffusion layer can be investigated in detail with the help of the developed model. The results of the simulations in this section are utilized for identifying the stress configuration on the gas diffusion layers for the evaluation of calculations made in following section. The use of computational methods assures the understanding of design and the dimensioning of fuel cell stack components e.g. endplates. Established model development in this section can be utilized for any application with similar properties.

4.3 Fuel Cell Simulations (Cross-Section Model)

In the previous section the fuel cell stack was investigated regarding the compression and thermomechanical characteristics. In order to analyze the effect of compression on the fuel cell performance, an electrochemical cross section model is established and presented in this section, which can be seen in Figure 4.1.

First a detailed literature survey of the electrochemical modeling of the fuel cells is performed and given in this section. The development of the FEM model consisting of the geometrical and physical conditioning is explained gradually in the next subsection. The computational fluid dynamics simulations are performed for the whole flow field structure without electrochemical modeling in order to get the boundary settings of the cross-section model. The calculated pressure values in flowfield simulations are utilized in the cross-section model as boundary conditions for the electrochemical modeling. The results of the fluid dynamics simulations are presented under boundary settings in regarding subsection.

The results and the evaluation of the electrochemical computations are given in the last subsection by parametrizing the compression force on the cross section model.

4.3.1 Literature Model Overview

The detailed literature survey of the electrochemical modeling of the PEM fuel cells with the help of discretization methods is handled in this subsection.

A detailed review of the electrochemical modeling of the fuel cells is given in [10, 18, 151-155]. Several electrochemical PEM fuel cell models have been published with different software packages [52, 53, 149, 156-196]. The modeling efforts can be classified from one-dimensional to 3 dimensional large-scale models. Due to the progress in computational capacity large-scale models have been developed more often recently [156-162]. Nevertheless several assumptions and simplifications are made for each modeling approach with regard to the focused cell properties. Some simulation models contain the water transport phenomenon as well [149, 156-158, 163-178]. Some 2-dimensional and 3-dimensional electrochemical models take the effect of mechanical compression on the cell performance into consideration [52, 53, 166-169, 174-184]. Some selected modeling efforts in the field of the high temperature PEM fuel cells are given in [194-196].

In this section a 3-dimensional cross-section model is developed. This 3-dimensional model contains electrochemistry and structural mechanics together and integrated with two additional computational models, which expands the evaluation and verification. The results of the thermomechanical stack simulations presented in previous section and flow field simulations presented in the following subsection are implemented into this electrochemical model as boundary and material properties. This has been achieved first time, which enables the exact analysis and verifications of the simulation results with measured data. This the first model containing two adjacent flow channels with parameterized compression force acting on it, which enables the explicit analysis of the compression force on the GDL properties and thus on the cell performance. The cross section of the flow field geometry is not simplified and first time built exact the same as the real injection molded bipolar plates. The definition of the modeling is explained in detail in following subsections.

The temperature distribution is not expected to be significantly different on the model due to the size of the cross section geometry. Therefore the model is investigated for isothermal operation. The electrochemical model is built with COMSOL Multiphysics® according to [197-199].

4.3.2 Model Development

The simulation model built with COMSOL Multiphysics® is described in the following subsections. The geometry of the computational model is given in subsection Model Set-up. The governing equations in corresponding domains and their interaction between each other are described in section "Definition of Physics". Boundary settings of the modeling are explained in the following subsection with the help of the problem statement and assumptions. The implementation of the mechanical properties into the electrochemical modeling is achieved with the help of the definition of material properties, which is explained in section 4.3.2.4. The interactions of the physical interfaces are dealt in detail in subsection 4.3.2.5 for a proper understanding. The meshing properties with the assigned solver settings are given in subsection 4.3.2.6.

4.3.2.1 Model Set-up

The cross-section model is built with the help of CAD tools. First the geometry of the gas flow path is derived with the help of the bipolar plate geometry. The negative of the bipolar plate is taken for the flow field simulations for the anode and the cathode respectively. Gas connector geometries are utilized to complete the geometry of the gas flow path. The exact representation of the gas flow paths in a standard ZBT fuel cell stack can be seen in Figure 4.36. The media follow the paths represented in Figure 4.36. This geometry is utilized for the fluid dynamics simulations. The fluid dynamics modeling properties for the whole flow paths are known via the measured data.

Modeling of the electrochemical behavior of fuel cells in stack size with numerical methods is not possible due to limited computational capacity. A cross-section model of a single fuel cell is established for the electrochemical calculations utilizing the geometry of the gas flow paths. The bipolar plates and the cell components are integrated into the derived cross-section model as illustrated in Figure 4.36. The results of the fluid dynamics simulations for the whole flow field are applied to the cross-section model to define the boundary settings as explained in 4.3.2.3.

The cross-section model is prepared for the implementation of the compression pressure. The outer faces of the bipolar plates are built appropriate for the application of the mechanical pressure on the cross-section model.

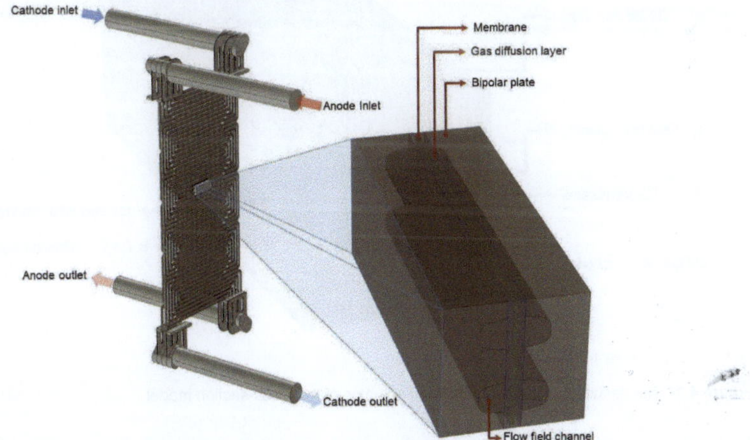

Figure 4.36 Flow field path geometry and derived cross-section model

Two adjacent flow channels are selected for the cross-section model in order to investigate the mechanical effect of the lands on the gas diffusion layers. The adjacent channels are chosen at the bending loop of the flow field structure in the middle of the bipolar plate as can be seen in Figure 4.36. This results in a certain pressure drop due to the bending corner in the flow field and facilitates the analysis of the gas diffusion through the GDL. The catalyst layers are neglected in the FEM modeling due to the selected electrochemical modeling approach and their spatial dimensions.

4.3.2.2 Definition of Physics

In order to simulate the PEM fuel cell performance under mechanical compression, the multiphysical cross-section model uses structural mechanics, current balances, electrochemical modeling, momentum and mass transports. Due to the small dimensions of the cross-section model, the effect of the temperature on the electrochemical modeling is not regarded here [197-199].

The cross-section model shown in Figure 4.36-37 consists of the computational domains, which are listed in Table 4.3 with the definition of the assigned governing equations for each domain. Each differential equation represents a corresponding physical phenomenon. The multiphysics model finds the solution of each governing differential equation for corresponding variables within the regarding computational domain. The solution of the differential equations for boundary value problems has already been explained in section 4.1.

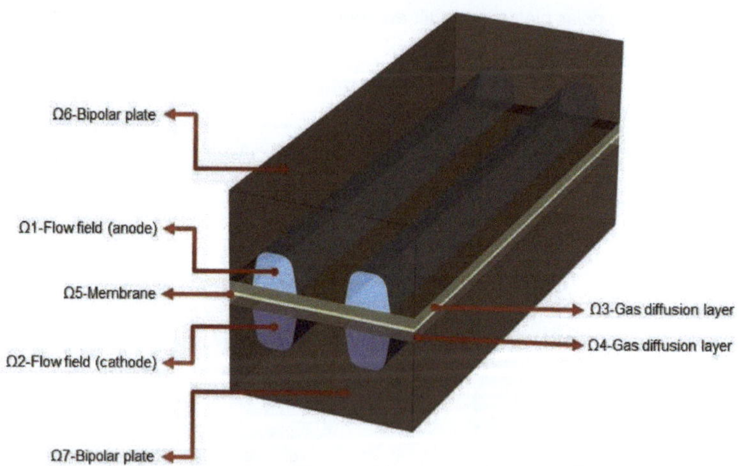

Figure 4.37 The definition of the computational domains of the cross-section model

The interaction of the different physical phenomena is defined with the help of model couplings, which is explained in section 4.3.2.5 in detail.

Table 4.3 The definitions of the physical modeling in the corresponding computational domains

Computational Domain	Assigned Governing Equations for Physical Modeling
Ω1, Ω2 - Flow field channels	Navier-Stokes differential equations for fluid dynamic calculations Maxwell-Stefan differential equations for transport of species
Ω3, Ω4 - Gas diffusion layers	Darcy's equation for flow in porous media Maxwell-Stefan differential equations for transport of species Ohm`s law for the charge transport Poison's differential equation for the structural mechanics
Ω5 - Membrane	Ohm`s law for the ion transport Poison's differential equation for the structural mechanics
Ω6, Ω7 - Bipolar plates	Ohm`s law for the charge transport Poison's differential equation for the structural mechanics

4.3.2.2.1 Momentum Transport

Navier-Stokes differential equations are assigned for the momentum transport in the flow field channels (Ω1, Ω2) for anode and cathode side respectively (see Eq. 4.32). Laminar and incompressible flow is assumed for the flow fields.

$$\rho \nabla \cdot \boldsymbol{u} = 0$$
$$\rho(\boldsymbol{u} \cdot \nabla)\boldsymbol{u} = \nabla \cdot \left(-p\boldsymbol{I} + \eta \cdot \left(\nabla \boldsymbol{u} + (\nabla \boldsymbol{u})^T\right)\right)$$

Eq. 4.32

The velocity field and pressure distribution of the gas flow channels are calculated by the computation of the velocity vector $\boldsymbol{u}$ and pressure tensor p respectively. η denotes the dynamic viscosity of the media. The density ρ of the media in anode and cathode side is calculated by ideal gas law as given in Eq. 4.33 according to the calculated mole fractions.

$$\rho = \frac{p}{R_g T} \sum_i M_i x_i$$

Eq. 4.33

x_i and M_i represent the mole fraction and molar mass of the species respectively. The index i stands for the regarding species. The mole fractions of the species (x_i) are coupled with the calculated values given in 4.3.2.2.2. The pressure p is computed and R_g represents the gas constant. The value of the temperature is predefined as mentioned before.

Darcy's law is utilized in the model (see Eq. 4.34), in order to calculate the gas flows in the porous diffusion layers (Ω3, Ω4).

$$\nabla \cdot (\rho \boldsymbol{u}) = 0$$
$$\boldsymbol{u} = -\frac{\kappa_p}{\eta} \nabla p$$

Eq. 4.34

The velocity field ($\boldsymbol{u}$) is calculated for the gas diffusion layers. κ_p represents the permeability of the porous medium.

4.3.2.2.2 Mass Transport

In order to calculate the mass transport in the model, the Maxwell-Stefan equations (see Eq. 4.35) are implemented into the corresponding subdomains ($\Omega1$, $\Omega2$, $\Omega3$, $\Omega4$). Two separate mass transport interfaces are assigned for the anode and cathode sides with the predefined species respectively.

$$\nabla \cdot \left[-\rho\omega_i \sum_{j=1}^{N} D_{ij} \left\{ \frac{M}{M_j}\left(\nabla\omega_j + \omega_j \frac{\nabla M}{M} \right) + (x_j + \omega_j)\frac{\nabla p}{p} \right\} + \omega_i \rho \boldsymbol{u} + D_i^T \frac{\nabla T}{T} \right] = R_i \qquad \text{Eq. 4.35}$$

The temperature term T vanishes due to the assumed constant operating temperature in the simulations. The velocity vector $\boldsymbol{u}$ is coupled with the momentum transport defined in the previous section. The calculated velocity vectors $\boldsymbol{u}$ for the gases in the gas diffusion layers and flow field paths are assigned to the related domains for the mass transport. The indices i and j stand for the regarding species in the model. Two species at the anode side (H_2, H_2O) and three species at the cathode side (O_2, H_2O, N_2) are considered in the model. The term D_{ij} denotes the Maxwell-Stefan diffusivity matrix and is computed with the help of the binary diffusivities between species. The effective binary diffusivities of the species are taken from [197]. The mass fraction term ω_i is calculated for each species in the mass transport physics interface. For instance two species (H_2, H_2O) are defined on the anode side of the fuel cell. The mass fraction variables for the species are ω_1 and ω_2 for the H_2 and H_2O respectively. The hydrogen mass fraction ω_1 is calculated as in Eq. 4.36. Mass fraction of the water is given in Eq. 4.37 due to the conservation of the mass.

$$\nabla \cdot \left\{ -\rho\omega_1 \sum_j \left[D_{1j} \left(\nabla x_j + (x_j - \omega_j)\left(\frac{\nabla p}{p} \right) \right) \right] \right\} = -(\rho \boldsymbol{u} \cdot \nabla \omega_1) \qquad \text{Eq. 4.36}$$

$$\omega_2 = 1 - \omega_1 \qquad \text{Eq. 4.37}$$

The calculation of the mass transport for the cathode side can be found analogous to anode side as given above.

4.3.2.2.3 Charge Transport

For the charge transport the following differential equation is assigned to the regarding domains in the model ($\Omega3$, $\Omega4$, $\Omega6$, $\Omega7$).

$$-\nabla \cdot (\varphi_s \cdot \nabla\phi_s) = -Q_s \qquad \text{Eq. 4.38}$$

ϕ_s is the electrical potential. φ_s represents the electrical conductivity of the components defined in material properties. Q_s is source or sink term respectively. Analogous to the electrical charge transport, ion transport through the membrane ($\Omega5$) is governed with the Eq. 4.39.

$$-\nabla \cdot (\kappa_m \cdot \nabla \phi_m) = -Q_m \qquad\qquad \text{Eq. 4.39}$$

ϕ_m, κ_m represent ion potential and ion conductivity respectively.

4.3.2.2.4 Structural Mechanics

In order to calculate the stress distribution and deformation on the cross section model Eq. 4.40 is implemented into the model for the corresponding domains ($\Omega 3$, $\Omega 4$, $\Omega 5$, $\Omega 6$, $\Omega 7$). The deformation calculated in structural mechanics is coupled with momentum and charge transport interfaces via the definition of material properties given in detail in section 4.3.2.4.

$$\nabla \cdot (D\nabla s) = 0 \qquad\qquad \text{Eq. 4.40}$$

s represents the deformation vector. The definition of the stiffness matrix D is given before in Eq. 4.20.

4.3.2.2.5 Electrochemical modeling

The basic reactions taking place in the PEM fuel cell are explained in section 2.2.2. The hydrogen consumption at the anode side produces protons, which travel through the membrane. In electrochemical modeling it is assumed that each travelling proton drags three water molecules. The resulting transport reaction occurring on the anode side is as given in Eq. 4.41.

$$H_2 + 2\lambda_{H_2O}H_2O \Rightarrow 2\left(H^+ + \lambda_{H_2O}H_2O\right) + 2e^- \qquad\qquad \text{Eq. 4.41}$$

λ_{H_2O} denotes the number of transported water molecules per proton. The oxygen molecules on the cathode side forms water together with the transported protons and electrons. The resulting reaction on the cathode side with the dragged water molecules is given in Eq. 4.42.

$$O_2 + 4\left(H^+ + \lambda_{H_2O}H_2O\right) + 4e^- \Rightarrow \left(2 + 4\lambda_{H_2O}\right)H_2O \qquad\qquad \text{Eq. 4.42}$$

The electrochemical reactions occur on the active catalyst layers as illustrated in Figure 2.5. The electrochemical modeling treats the catalyst layers as boundaries. This is called as agglomerate modeling [197].The electrochemical modeling bases on the boundary settings, which are given in detail in section 4.3.2.3. The reactive layers are taken into consideration in the charge and mass transport in the electrochemical modeling. The interaction between the physical interfaces is established with the help of boundary and domain settings together with the material definition. The detailed illustration of the model couplings is given in section 4.3.2.5 for a better understanding.

Following assumptions are made for the multiphysical modeling of cross-section geometry:

- Steady state analysis with stationary conditions is considered.
- Laminar and incompressible flow are taken for the momentum transport due to the low pressure and velocity profiles.
- Ideal gas mixture is assumed.
- Isothermal conditions are assigned for the cross section model.

- Two phase-transport is not regarded in the modeling (water exists only in gas phase).
- The membrane is assumed to be fully humidified.
- The membrane swelling effect is not considered in the simulations (as realized by the measurements presented in section 3.1.1.1).
- Isotropic permeability is used for the gas diffusion layers.
- Interaction of the flow channels with the gas diffusion layers are provided by implementation of the pressure values.
- Catalyst layers are neglected regarding the properties of the electrochemical modeling.

4.3.2.3 Boundary Settings

The assigned boundary settings and their governing equations are explained in following subsections for each physics interface separately.

4.3.2.3.1 Momentum Transport

In order to find the boundary settings for the momentum transport of the cross-section model, the whole flow path has to be simulated from fluid dynamics point of view as explained before. The required boundary conditions for the momentum transport of the cross section model are provided by the utilization of the flow field simulation results presented in this section. The gas flow path of the standard ZBT bipolar plate design is depicted in Figure 4.36. The flow field geometry is sectioned precisely to extract the boundary settings for the cross section model as in Figure 4.36. The boundary settings for the flow field simulations are acquired by the stack measurements. The flow field geometries and assigned constraints for the flow field simulations can be seen in Figure 4.38.

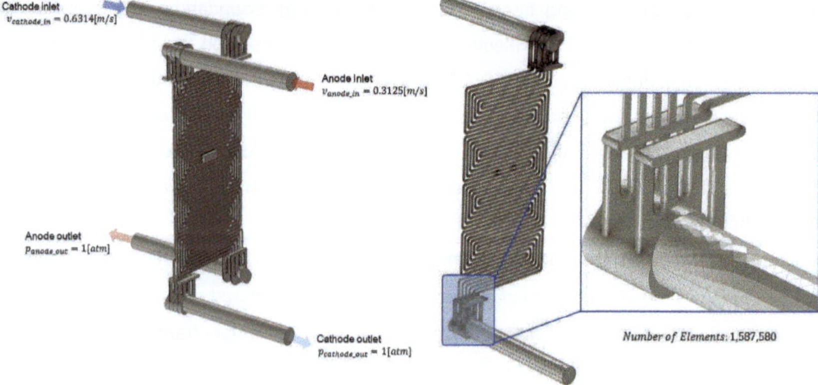

Figure 4.38 The constraints for the flow field simulations   **Figure 4.39** Meshing of the flow field geometry

The meshing of the flow field geometry and mesh properties are given in Figure 4.39. The velocity profile in the flow field of the anode side (hydrogen) is depicted in Figure 4.40-41. The streamline illustration of the velocity field on the fuel cell manifolds and flow field inlet is given in Figure 4.41 for the analysis of the manifolds. The hydrogen flow from manifolds into the flow field channels can be seen in Figure 4.41.

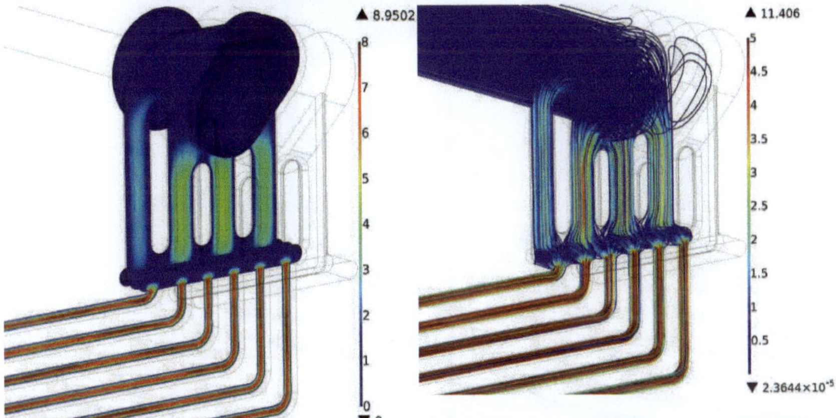

Figure 4.40 The velocity field [m/s] on the selected slices for anode

Figure 4.41 Streamline plot of the velocity field [m/s] for anode

The pressure distribution on the whole flow field path for the anode side can be seen in Figure 4.42. The pressure development on the cross section geometry is depicted in Figure 4.43, which is integrated into the cross section model as boundary settings for the electrochemical modeling.

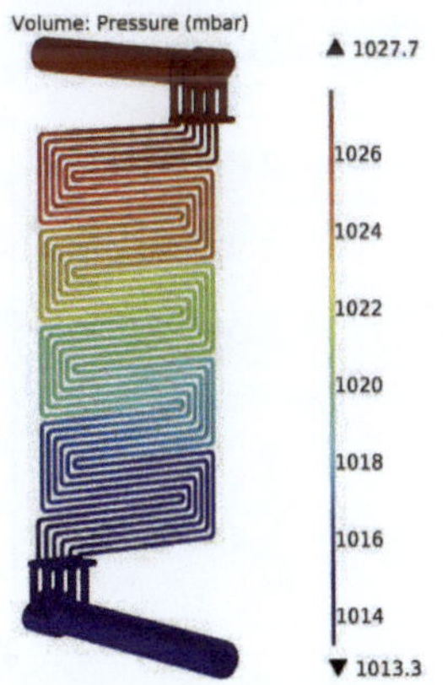

Figure 4.42 The pressure distribution of the anode

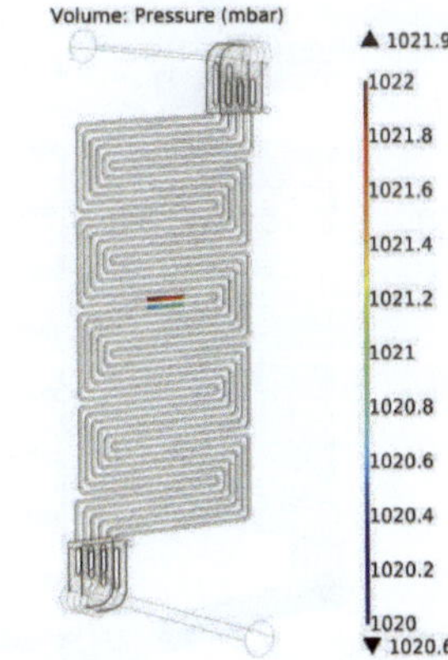

Figure 4.43 The pressure distribution of cross section model for anode

The velocity profile in the cathode side (air) of the fuel cell is depicted in Figure 4.44-45 analogous to the anode side. The flow of the air into the flow field channels can be seen in Figure 4.45. The high volumetric flow of the air on the cathode side results in high velocity profile in the flow field.

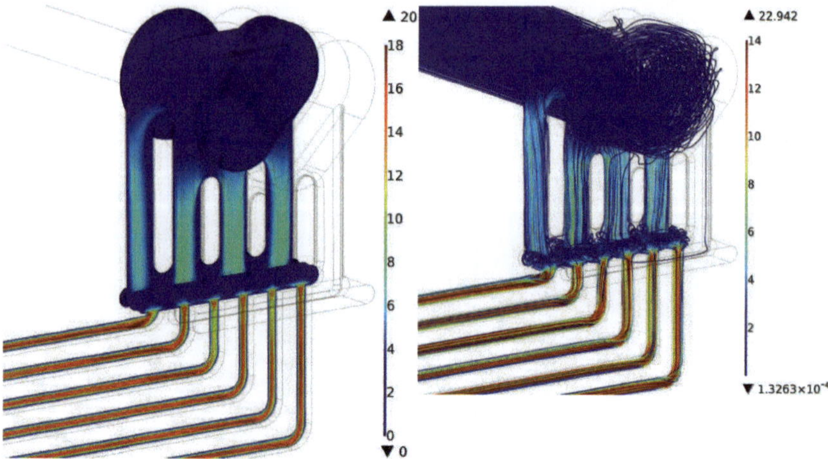

Figure 4.44 The velocity field [m/s] on the selected slices for cathode

Figure 4.45 Streamline plot of the velocity field [m/s] for cathode

The pressure distribution on the whole flow field path for the cathode side of the fuel cell is shown in Figure 4.46. The pressure development on the cross section geometry of the cathode side is given in Figure 4.47. The analysis of the pressure drop on the flow path is essential for the flow field design.

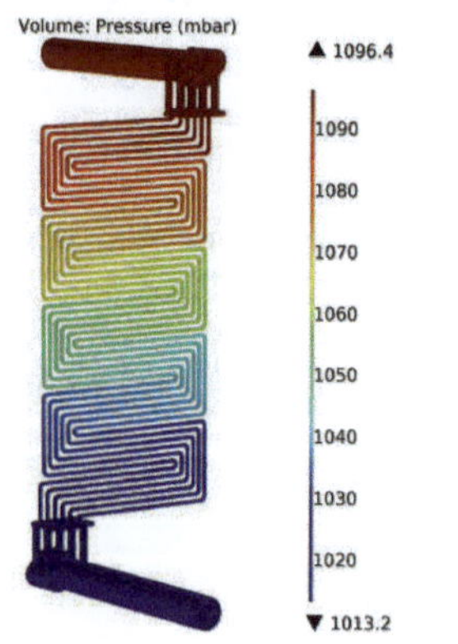

Figure 4.46 The pressure distribution [mbar] of the anode

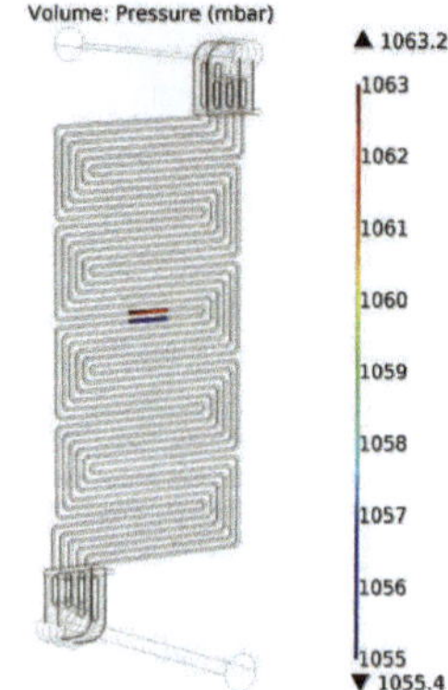

Figure 4.47 The pressure distribution [mbar] of cross section model for anode

The fluid dynamical simulations of the flow fields do not only deliver the boundary settings for the cross section model but also enable the analysis of the flow field design. The distribution of the pressure and the velocity field for the anode and cathode side are essential to analyze the flow field design [202, 203]. An even distributed pressure profile is required in order to transport the produced water from the fuel cell and provide a proper gas supply.

The simulated pressure values on the cutting surfaces of the flow field geometry are integrated into the momentum transport of the cross section model as boundary settings (see Figure 4.48) for anode and cathode side respectively. The coupling between the momentum transport for the flow field and the porous medium flow in gas diffusion layers is achieved by integration of the calculated pressure value into the porous medium flow interface as boundary conditions. The calculated pressure profile on the cross section model is implemented into the gas diffusion layers as boundary conditions on the neighboring surfaces as marked in green in the Figure 4.49. The catalyst layers of the anode and cathode side are regarded as boundaries as explained before. Boundary settings are assigned for the porous medium flow on the selected boundaries as highlighted in the Figure 4.49. The boundaries marked in blue refer to the catalyst layers for the anode and cathode side respectively. The equations acting on these boundaries are given in Eq. 4.43-44 representing the gas flow of the reactants and products due to the electrochemical reactions. These are coupled with the simultaneous calculated current density on the membrane due to the electrochemical modeling.

$$-n \cdot u_{anode} = \frac{J_{anode}}{\rho F}\left(\frac{M_{H_2}}{2} + \lambda_{H_2O}M_{H_2O}\right) \qquad\qquad \text{Eq. 4.43}$$

$$-n \cdot u_{cathode} = \frac{J_{cathode}}{\rho F}\left(\frac{M_{O_2}}{4} + \left(\frac{1}{2} + \lambda_{H_2O}\right)M_{H_2O}\right) \qquad\qquad \text{Eq. 4.44}$$

n is the normal vector which represents the boundary. u_i and J_i represent the normal inlet velocity and current density for anode and cathode side respectively.

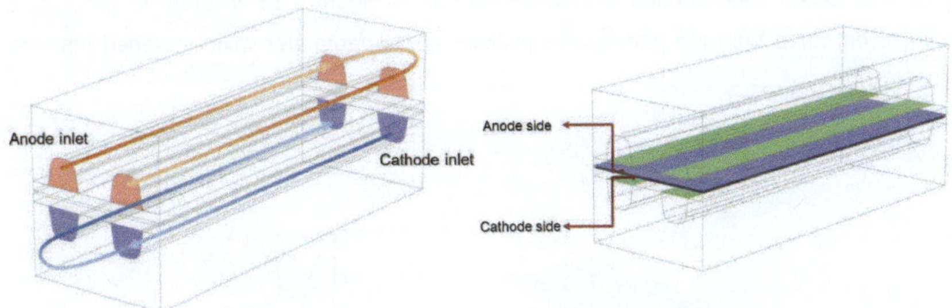

Figure 4.48 Boundary couplings for the momentum transport

Figure 4.49 Boundary couplings for the momentum transport in porous medium

Wall and symmetry conditions are assigned appropriately to the remaining boundaries of the momentum transports for the flow fields and gas diffusion layers. The calculated pressure profile on the flow field channels and gas diffusion layers are coupled with the other physical interfaces.

4.3.2.3.2 Mass Transport

The inlet and outlet conditions are assigned to the mass transport analogous to the momentum transport. Different from the momentum transport, mass transport handles the flow field channels and gas diffusion layers together for the anode and cathode side respectively. Mass fractions of the regarding species for the inlets are applied on the boundaries as illustrated in Figure 4.48. The mass fraction values are acquired by the measurements and assumed to be constant at the inlets for the cross section model. The reactions acting on the catalyst layers are defined on the boundaries as depicted in Figure 4.49 analogous to the momentum transport. Calculated current density in the electrochemical modeling is integrated into the mass transport as boundary settings on the membrane surfaces as defined in Eq. 4.45-47 for the anode and cathode sides respectively.

$$-n \cdot N_{H_2}\big|_{anode} = \frac{J_{anode}}{2F} M_{H_2} \qquad\qquad \text{Eq. 4.45}$$

$$-n \cdot N_{O_2}\big|_{cathode} = \frac{J_{cathode}}{4F} M_{O_2} \qquad\qquad \text{Eq. 4.46}$$

$$n \cdot N_{H_2O}\big|_{cathode} = \frac{J_{cathode}}{F}\left(\frac{1}{2} + \lambda_{H_2O}\right) M_{H_2O} \qquad\qquad \text{Eq. 4.47}$$

The notation is same as explained in previous section. Wall and symmetry conditions are applied appropriately to the remaining boundaries of the mass transport.

4.3.2.3.3 Charge Transport

The boundary conditions for the charge transport are highlighted in Figure 4.50. The electric potential for the anode is assumed to have a value of 0 [V]. The cell potential is assigned as a parameter to the outer surface of cathode side as in Figure 4.50. In order to generate a polarization curve from the performed simulations, this parameter value is varied between 0.01-1 [V].

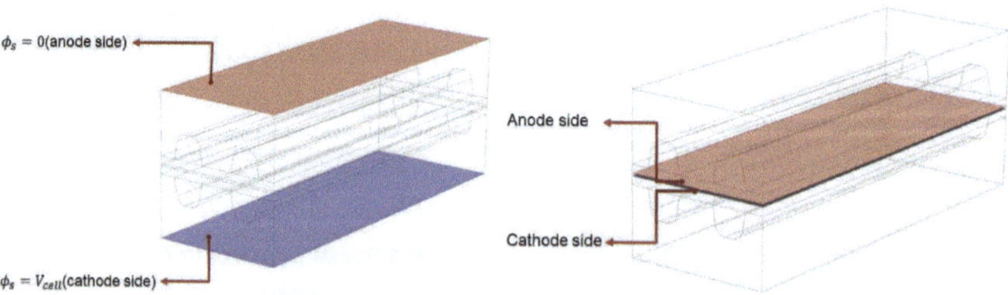

Figure 4.50 Boundary couplings for the charge transport

Figure 4.51 Boundary couplings for the ion transport

Calculated current densities in electrochemical modeling are implemented into the charge transport as boundary settings for the membrane contacting surfaces of the anode and cathode side as depicted in Figure 4.51.

The contact impedance is applied on the contacting surfaces between the bipolar plates and gas diffusion layers. Governing equations for the contact impedance boundaries are given in Eq. 4.48. The indices 1 and 2 refer to the bipolar plate and gas diffusion sides of the contacting boundary.

$$n \cdot J_1 = \frac{1}{\rho_s}(V_1 - V_2)$$

$$n \cdot J_2 = \frac{1}{\rho_s}(V_2 - V_1)$$

Eq. 4.48

J and ρ_s are the current density and surface resistance respectively. V is calculated potential variable. Surface resistance ρ_s [$\Omega \cdot m2$] is defined as a function of the mechanical compression, which is handled in section 4.3.2.4. This contributes to the interaction of the charge transport with the structural mechanics interface held in next section. Electrical insulation is assigned to the remaining surfaces as boundary condition for the charge transport. Boundary settings for the ion transport are established analogous to the charge transport. Calculated current densities for the anode and cathode side in the electrochemical modeling are applied on the membrane surfaces as depicted in Figure 4.51. Ionic insulation is set as boundary condition for the remaining surfaces of the membrane for ion transport.

4.3.2.3.4 Structural Mechanics

The boundary conditions depicted in Figure 4.52 and Figure 4.53 are applied for the structural mechanics interface. The boundary load is assigned for one side of the cross section model and set as a parameter. This is varied by the simulations in order to analyze the effect of mechanical compression on the cell performance.

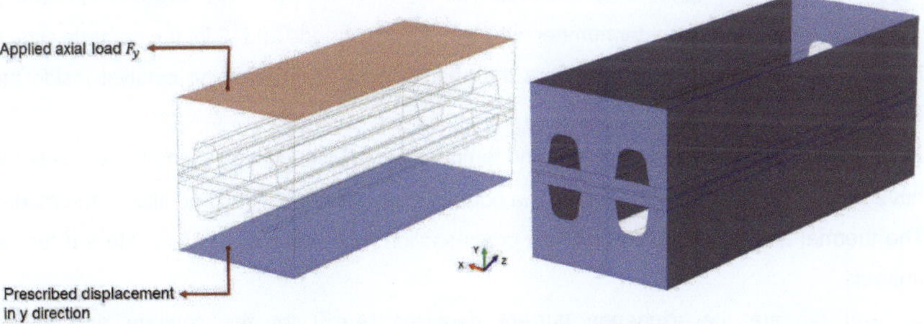

Figure 4.52 Boundary settings for the structural mechanics

Figure 4.53 Symmetry conditions for the structural mechanics

The in blue highlighted surface in Figure 4.52 is prescribed to have a displacement value of 0[m] in y direction. The remaining cutting surfaces of the cross-section model are held as symmetry boundaries.

4.3.2.3.5 Electrochemical modeling

The boundary settings build the agglomerate modeling of the electrochemical reactions in the cross-section model. The catalyst layers are regarded as boundaries in the electrochemical agglomerate modeling as depicted in Figure 4.51.

The current densities on the cathode and anode side are calculated by Butler-Volmer approach according to [197-199]. Eq. 4.49-50 are assigned to the boundaries of the anode and cathode side respectively. The indices a and c represent the anode and cathode correspondingly.

$$i_a = L_{act}(1 - \varepsilon_{mac})J_{agg,a} \qquad\qquad \text{Eq. 4.49}$$

$$i_c = L_{act}(1 - \varepsilon_{mac})J_{agg,c} \qquad\qquad \text{Eq. 4.50}$$

L_{act} and ε_{mac} represent the active layer's thickness [m] and macroscopic porosity respectively. J_{agg} is the current density which is described in the following equations for the anode and cathode side.

$$J_{agg,c} = 6n_c F \left(\frac{D_{agg}}{R_{agg}}\right)(1 - \beta_c \coth\beta_c)c_{O_2,agg} \qquad\qquad \text{Eq. 4.51}$$

$$J_{agg,a} = 6n_a F \left(\frac{D_{agg}}{R_{agg}}\right)(1 - \beta_a \coth\beta_a)\left[c_{H_2,agg} - c_{H_2,ref}exp\left(\frac{-2F}{R_g T}\eta_a\right)\right] \qquad\qquad \text{Eq. 4.52}$$

β_a and β_c are defined in Eq. 4.53 and Eq. 4.54.

$$\beta_a = \sqrt{\frac{i_{0a}S R_{agg}^2}{2F c_{H_2,ref} D_{agg}}} \qquad\qquad \text{Eq. 4.53}$$

$$\beta_c = \sqrt{\frac{i_{0c}S R_{agg}^2}{4F c_{O_2,ref} D_{agg}}}\, exp\left(\frac{-F}{2R_g T}\eta_c\right) \qquad\qquad \text{Eq. 4.54}$$

R_{agg} and D_{agg} are agglomerate radius [m] and gas diffusivity [m²/s] respectively. n_c and n_a stand for the charge transfer number, which is 1 for the anode and -2 for the cathode side. F is the Faraday's constant [C/mol]. S is the specific surface area of the catalyst inside the agglomerate.

R_g and T are gas constant [J/(mol·K)] and temperature [K] respectively. Temperature is set to have a constant value of 70°C due to the point of interest of this study and size of the model. The thermal analysis is built within the cross-section model in order to facilitate the further analysis.

i_{0c} and i_{0a} are the exchange current densities [A/m²] for the cathode and anode correspondingly. $c_{O_2,ref}$ and $c_{H_2,ref}$ stand for the reference molar concentrations on the related agglomerate surface [mol/m³]. The overvoltage η_i is given in Eq. 4.55 and Eq. 4.56 for the anode and cathode respectively.

$$\eta_a = \phi_s - \phi_m - E_{eq,a} \qquad\qquad \text{Eq. 4.55}$$

$$\eta_c = \phi_s - \phi_m - E_{eq,c} \qquad\qquad \text{Eq. 4.56}$$

E_{eq} is the equilibrium voltage [V]. ϕ_s and ϕ_m are the electrical and ionic potential variables, which are coupled with the charge and ion transport interfaces. c_{agg,H_2} and c_{agg,O_2} are hydrogen and oxygen concentrations on the agglomerate surfaces respectively and defined with the help of the Henry's law as given in Eq. 4.57 and Eq. 4.58.

$$c_{agg,H_2} = \frac{p_{H_2} x_{H_2}}{K_{H_2}} \qquad\qquad \text{Eq. 4.57}$$

$$c_{agg,O_2} = \frac{p_{O_2} x_{O_2}}{K_{O_2}} \qquad\qquad \text{Eq. 4.58}$$

K_{H_2} and K_{O_2} are Henry's gas constants [Pa·m³/mol] for hydrogen and oxygen correspondingly. p_{H_2} and p_{O_2} are the pressure values on the anode and cathode side of the agglomerate, which are coupled with the calculated pressure values within the porous medium flow interface. x_{H_2} and x_{O_2} are the calculated mole fraction values in the mass transport interfaces for the hydrogen and oxygen respectively. This is utilized to find the partial pressure of the regarding species on the agglomerate surfaces. The definition and the values of the parameters are given in Table 4.4 [197].

Table 4.4 Parameter list for the electrochemical modeling

Parameter	Description	Unit	Value
L_{act}	Active layer thickness	[μm]	10
ε_{mac}	Macroscopic porosity between agglomerates		0.4
R_{agg}	Agglomerate radius	[μm]	0.1
n_c	Charge transfer number		1
n_a	Charge transfer number		-2
F	Faraday's constant	[C/mol].	96485
R_g	Gas constant	[J/mol.K]	8.314
i_{0c}	Exchange current densities for cathode	[A/m²]	1
i_{0a}	Exchange current densities for anode	[A/m²]	1×10^5
$c_{O2,ref}$	Reference concentration of the oxygen	[mol/m³]	0.42688
$c_{H2,ref}$	Reference concentration of the hydrogen	[mol/m³]	1.2987
S	Specific surface area of the catalyst	[1/m]	1×10^7
$E_{eq,a}$	Equilibrium potential for anode	[V]	0
$E_{eq,c}$	Equilibrium potential for cathode	[V]	1
K_{H_2}	Henry's gas constants for hydrogen	[Pa·m³/mol]	3.9×10^4
K_{O_2}	Henry's gas constants for air	[Pa·m³/mol]	3.2×10^4
λ_{H_2O}	Water drag coefficient		3

4.3.2.4 Material Properties

Mechanical properties of the regarding components of the cross-section model are defined as in Table 4.1. Remaining required material properties of the components are given in Table 4.5.

Table 4.5 Material properties for the electrochemical cross-section model

Material Properties		Unit	Value	Source
BPP-Graphite-Compound				
	Electrical conductivity	[S/m]	1.1×10^4	[135]
Nafion-N-112				
	Ion conductivity	[S/m]	$\kappa(\lambda, T)$	[10]
GDL				
	Electrical conductivity	[S/m]	$\varphi_{ip}(\theta),\ \varphi_{tp}(\theta)$	[200]
	Permeability	[m²]	$\gamma(\theta)$	[200]
	Porosity		0.8	[145]

Ion conductivity of the membrane $\kappa(\lambda, T)$ is given in a function of the λ (water content) and T (temperature) as in Eq. 4.59.

$$\kappa = (0.005139\lambda - 0.00326)e^{\left[1268\left(\frac{1}{303} - \frac{1}{T}\right)\right]} \qquad \text{Eq. 4.59}$$

λ is set to have a constant value of 14 assuming hydrated membrane [10]. Modeling of the material properties enables the use of the cross-section model for further purposes. The material properties of the cell components depend highly on each other also on operating parameters. Several material properties of different cell components are dependent on the mechanical pressure, which affects the cell performance [53, 54, 177, 200, 201]. Electrical conductivity of the gas diffusion layers depends on the mechanical compression and differs in material directions. Hence electrical conductivity (φ_{ip}, φ_{tp} in and through plane directions respectively) is given as a function of compression factor θ. Permeability $\gamma(\theta)$ of gas diffusion layers is defined as a function of the compression factor θ analogous to the electrical conductivity. Contact resistance ρ_s between bipolar plates and gas diffusion layers is described with the same approach [200]. Compression factor θ is calculated in the structural mechanics module, which fulfills the interaction of the physical interfaces. The distribution of the mechanical compression differs on the cross-section model due to the geometry and material properties. This leads to geometrical analysis of the fuel cell performance from mechanical point of view. Varying the acting clamping force on the cross-section model changes the mechanical compression and corresponding material properties. This enables to analyze the effect of mechanical compression on cell performance, which is the point of interest of this study.

4.3.2.5 Interaction of the Physical Interfaces

The multiphysical cross-section model is built as explained in previous sections. The interaction of the physical interfaces is realized with the help of the domain settings, boundary conditions and material properties. For a better understanding of the interaction of the physical interfaces, the detailed illustration of the modeling couplings is given in Figure 4.54.

Figure 4.54 Interactions of the physical interfaces

Seven different physical interfaces are implemented into the multiphysical cross-section model. This refers to seven differential equations calculated together within the cross-section model.

The definition of the computational domains and the assigned physical interfaces are dealt in section 4.3.2.2. Momentum transport for the flow field channels delivers the velocity and pressure profile for mass transports in channels and for momentum transport in the gas diffusion layers. The momentum transport for the porous medium flow provides the pressure profile for electrochemical modeling and velocity profile for the mass transport. The calculated mass fractions of the species in anode and cathode are utilized for the calculation of the density profile and implemented into the electrochemical modeling in order to compute the current density. Ionic charge transport is coupled with the electrochemical modeling. Electrical potentials calculated in charge transport are integrated into the electrochemical modeling. As an objective of this study, the calculated compression factor in the structural mechanics interface is integrated into the material modeling and coupled with momentum and charge transports. The calculated current density values in electrochemical modeling are implemented into the momentum transport of the porous medium flow as boundary settings. The current density is integrated into the charge transports and mass transports, which fulfills the electrochemical multiphysics modeling. Within the multiphysical cross-section model the physical interfaces are integrated each other and solved simultaneously.

4.3.2.6 Meshing and Solver Settings

Meshing of the cross-section model is performed considering the geometrical and physical properties. The finer mesh requirement for the mass and momentum transport in comparison with structural mechanics and charge transport is regarded by meshing of the cross-section model. The meshing of the critical domains and boundaries is made more precisely due to the requirements of the electrochemical modeling. The computational capacity is considered by assuring the independence of the results from the mesh quality. Mesh distribution is provided for a proper convergence of the solution. Mapped, swept and free meshing options are utilized to build the meshing of the FEM model. Mesh elements of the cross-section model can be seen in Figure 4.55. The mesh of the cross-section model consists of 135,496 elements with an element quality of 0.247.

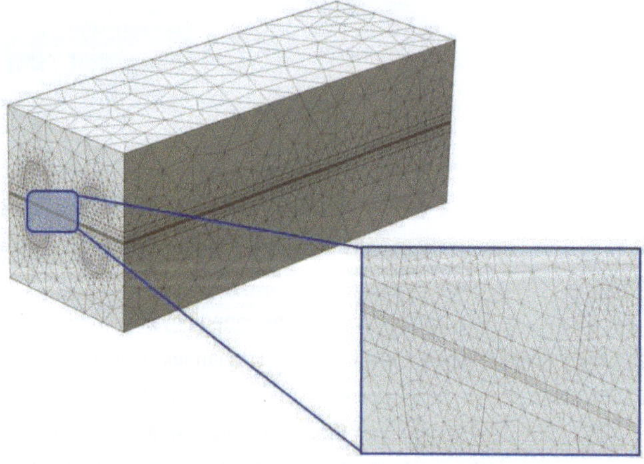

Figure 4.55 Meshing of the cross-section model

Solver settings and initial conditions are to be selected precisely due to the multiphysical properties of the cross-section model. A coupled solution setting with PARDISO Solver is assigned for each physical interface for the computation of the dependent variables.

An initial simulation with constant velocity values u_{anode} and $u_{cathode}$ for the momentum transport is performed. Using these simulation results as initial conditions for the parametric studies assures a certain convergence profile for the solution.

4.3.3 Results and Evaluation

The simulations based on the modeling approaches and assumptions explained before deliver supplementary information to the simulations presented in section 4.2. The large-scale stack simulations assure the analysis of internal stress distribution and deformation in stack size. In accordance with these results the simulations performed in this section provide complementary knowledge about the effect of mechanical pressure on fuel cell performance. It ensures the electrochemical analysis of the fuel cell for a cross-section geometry and assists to understand the effect of the mechanical clamping pressure on cell components.

A parametric study is performed varying the compression pressure and potential on the cathode side as explained before. The average Von Mises stress on the gas diffusion layers is calculated and compared with the simulation results performed in section 4.2. The equivalent compression pressure acting on the cross-section model is attained by the evaluation of the average Von Mises stress on the gas diffusion layers in stack size. An approximated equivalent compression pressure value of 4.5[MPa] is considered for the cross-section model representing the real stack clamping conditions including the thermal stress, which is defined with the help of the simulation results in section 4.2. Regarding the normal stack operation conditions, a compression pressure value of 4.5[MPa] and a voltage value of 0.6[V] for the cathode are taken for the results given below. The results are given with the same view. This ensures a proper coherence for comparison purposes.

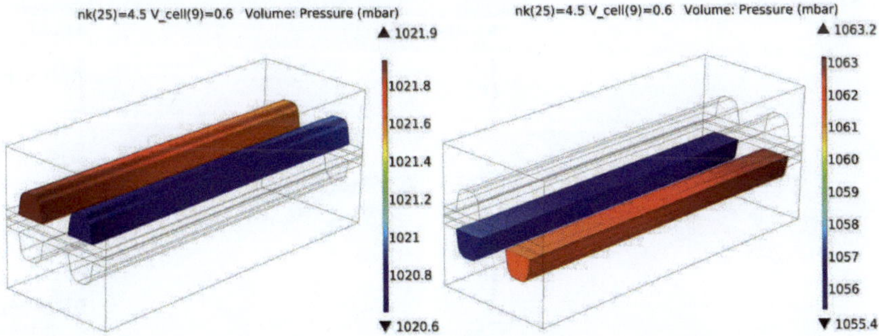

Figure 4.56 The pressure distribution [mbar] in 2 adjacent channels (anode)

Figure 4.57 The pressure distribution [mbar] in 2 adjacent channels (cathode)

Pressure distribution in the flow field channels can be seen in Figure 4.56 and Figure 4.57 for anode and cathode respectively. The difference in pressure values on the adjacent channels is developed by pressure drop due to the bending on the flow field. Pressure drop in a single channel is about 0.3[mbar] and 1.6[mbar] for anode and cathode respectively. Pressure values on the cathode and anode differ from each other due to the boundary settings for momentum transport. Higher pressure values on the cathode side results in higher pressure drop values as expected. Pressure distribution in the gas diffusion layers are depicted in Figure 4.58 and Figure 4.59 for the anode and cathode correspondingly.

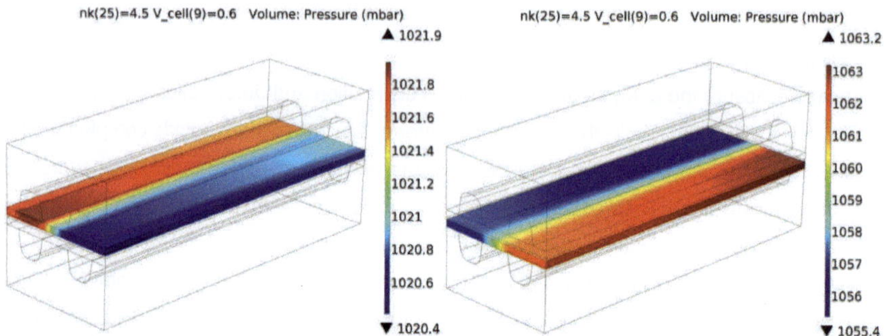

Figure 4.58 The pressure distribution [mbar] on the gas diffusion layer of the anode

Figure 4.59 The pressure distribution [mbar] on the gas diffusion layer of cathode

Gas flow though the gas diffusion layers is developed due to the pressure difference in adjacent flow channels. The flow through the gas diffusion layers plays a significant role for the cell performance. Gas diffusion layers provide the distribution of the reactants in order to utilize the whole active area of the membrane. Flow development though the porous medium can be analyzed together with the gas pressure distribution of the flow field channels. Simulated velocity profiles in the cross section model are given in Figure 4.60 and Figure 4.61 for the gas flow channels and gas diffusion layers respectively.

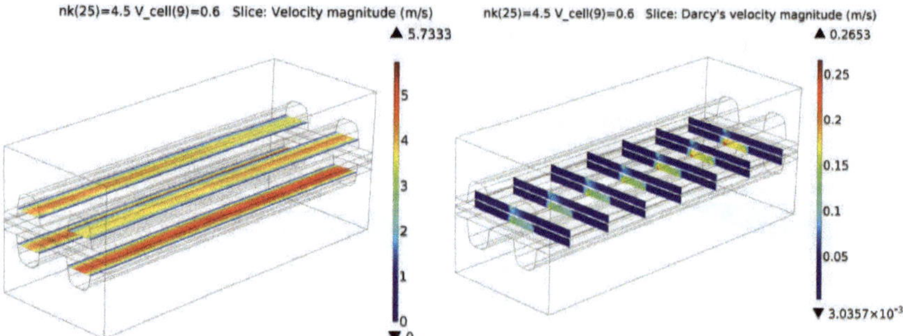

Figure 4.60 The velocity profile [m/s] on the slices in the middle of the flow channels

Figure 4.61 The velocity profile [m/s] on seven slices with equal intervals in gas diffusion layers

The velocity profile in the flow field channels is influenced mainly by the boundary settings of the momentum transport. Evaluating the velocity profile in the gas diffusion layers together with the pressure distribution given in Figure 4.58 and Figure 4.59 enables the proper analysis of the flow in the gas diffusion layers. Arrow lines of the velocity profile [m/s] in the middle of the gas diffusion layers are illustrated in Figure 4.62, in order to investigate the distribution of the velocity field. Flow directions through the gas diffusion layers can be noticed in Figure 4.62.

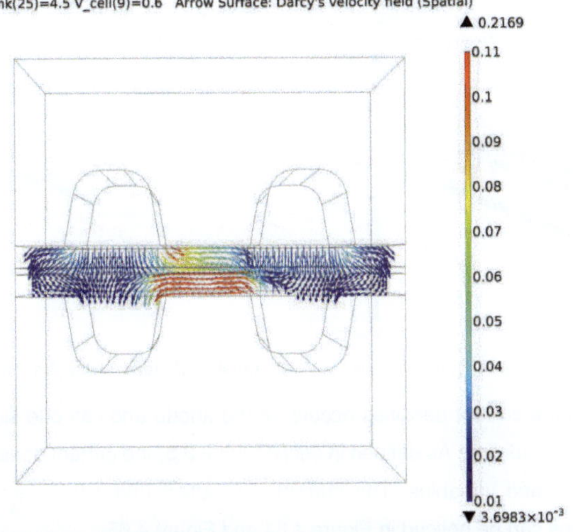

Figure 4.62 Arrow illustration of the Velocity profile[m/s] on the gas diffusion layers

The consumption of the hydrogen on the upper side (anode) and the production of the water on the cathode side can be realized. The flow between adjacent channels is induced by pressure difference as explained before. Permeability of the gas diffusion layers as a function of the compression pressure and the medium transport due to the reactions occurring on the membrane modify the pressure and the velocity development in gas diffusion layers. For a better understanding the permeability of the gas diffusion layers is depicted in Figure 4.63.

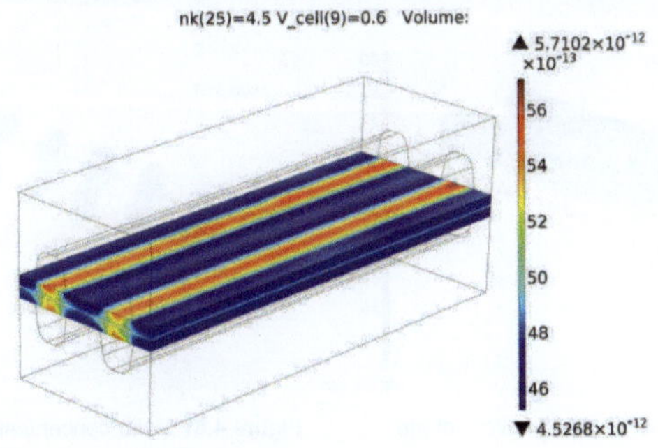

Figure 4.63 Permeability [m²] of the gas diffusion layers

Flow channels and bipolar plates contacting regions of the gas diffusion layers are under different mechanical load, which results in a change of the permeability values as it can be seen in Figure 4.63 and contributes to a change in the flow development of the gas diffusion layers. This enables to analyze the effect of mechanical compression on the flow development and cell performance as an objective of this study.

The current densities of the anode and cathode are given in Figure 4.64 and Figure 4.65 respectively.

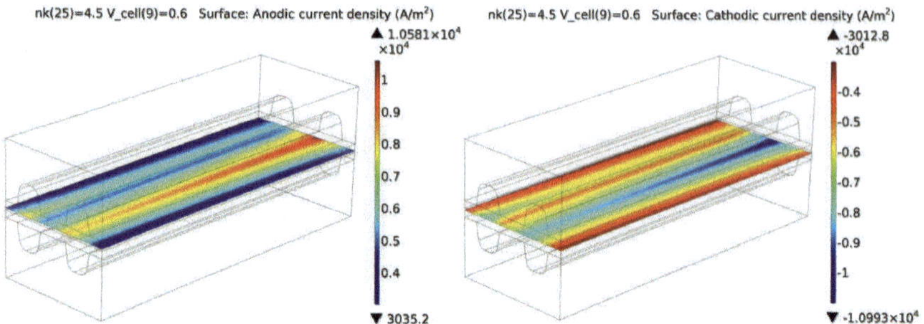

Figure 4.64 Current density [A/m²] on the anode   **Figure 4.65** Current density [A/m²] on the cathode

The distribution of the current densities occurs on the anode and cathode side according to the electrochemical modeling. As defined in section 4.3.2.3.5, the current densities depend on several parameters and variables. The current distribution under flow field channels and bipolar plate regions can be noticed in Figure 4.64 and Figure 4.65.

The molar concentrations of the species on the anode and cathode side of the fuel cell are essential and can be figured out with the following results. The results are illustrated for the whole domains together with seven selected slices for a proper analysis, which divide the domain in equal parts. The direction of the medium flow is depicted in the results in order to facilitate the analysis. The molar concentration of the hydrogen can be seen in Figure 4.66 and Figure 4.67.

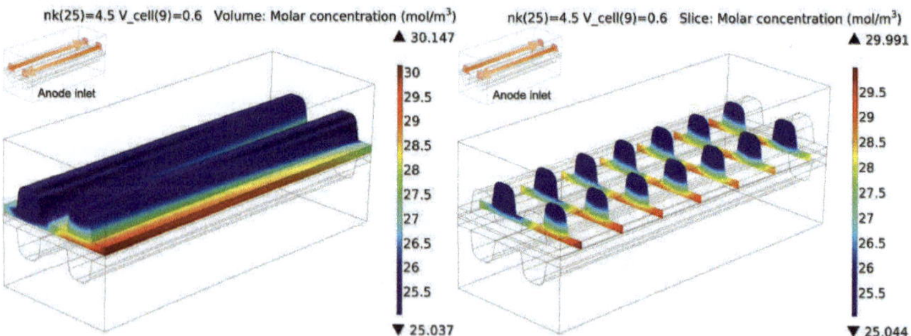

Figure 4.66 Molar concentration [mol/m³] of the hydrogen   **Figure 4.67** Molar concentration [mol/m³] of the hydrogen in slices

As it can be noticed in Figure 4.66 and Figure 4.67, the hydrogen molar concentration increases on the reacting layer of the membrane and through the flow direction. This is the result of the electroosmotic drag of water through the membrane as explained in electrochemical modeling. Water molecules are carried together with the traveling protons, which results in a decrease of the molar concentration of the water in the anode side. This results in a proportional increase of the molar concentration of the hydrogen, which is higher than the consumption of hydrogen. The distribution of the hydrogen molar concentration is developed mainly due to the electrochemical modeling. Molar concentration of water on the anode side is presented in Figure 4.68 and Figure 4.69 in accordance with the molar concentration development of hydrogen.

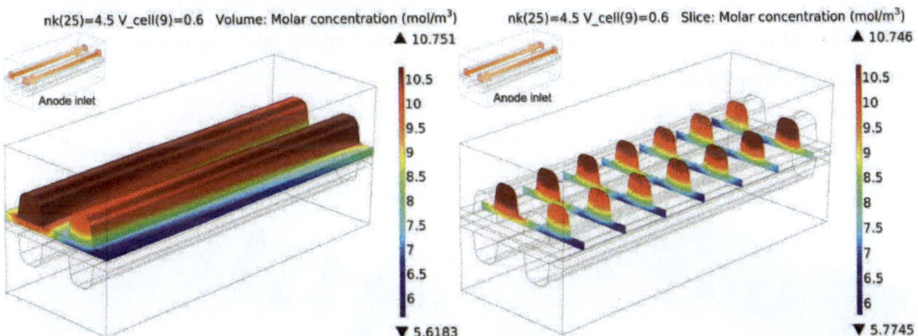

Figure 4.68 Molar concentration [mol/m³] of the water on the anode   **Figure 4.69** Molar concentration [mol/m³] of the water on the anode in slices

The molar concentration of the water on the anode is dependent on the hydrogen consumption and water drag as explained. The membrane is assumed to be fully humidified and has a constant proton conductivity in the performed simulations. This makes the fuel cell performance independent from the water content in the membrane.

Molar concentration of oxygen on the cathode side is depicted in Figure 4.70 and Figure 4.71 by adapting the view together with the flow direction analogous to the molar concentration of the species on the anode. Distribution of the oxygen plays a significant role on the performance of the fuel cell [10, 72]. The consumption of the oxygen on the cathode side can be noticed in Figure 4.70 and Figure 4.71 by following the molar concentration values through the flow field channels. Reactions are taking place intensely under bipolar lands neighboring the flow field channels. This results in a change in oxygen and water concentration values due to the oxygen consumption and water production respectively, which can be noticed in Figure 4.71 in the enclosed region with dashed line. The distribution of the oxygen molar concentration on the cutting surfaces is developed due to the boundary settings for the charge transports. Electrical and ionic insulation are assigned to the cutting surfaces of the cross section model respectively.

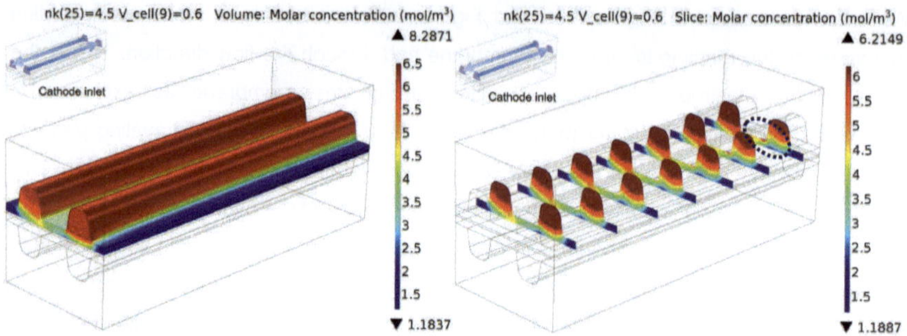

Figure 4.70 Molar concentration [mol/m³] of the oxygen

Figure 4.71 Molar concentration [mol/m³] of the oxygen in slices

The water production on the cathode due to the reactions taking place on the membrane can be noticed in Figure 4.72 and Figure 4.73.

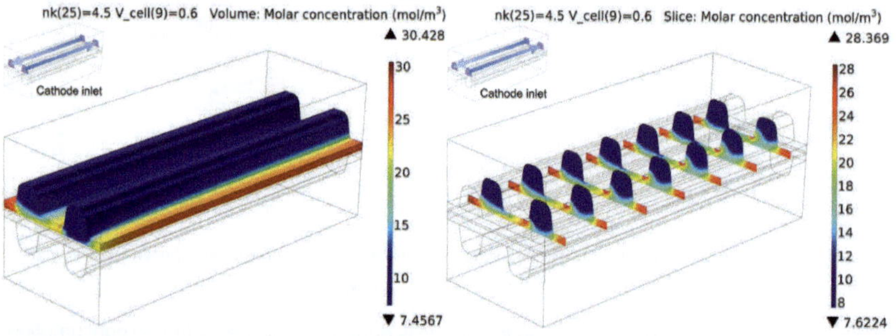

Figure 4.72 Molar concentration [mol/m³] of the water on the cathode

Figure 4.73 Molar concentration [mol/m³] of the water on the cathode in slices

Molar concentration of the water on the cathode is developed due to the boundary settings for the mass transport and produced water. The transport of the water molecules from the anode to the cathode side is also considered as explained. Molar concentration of nitrogen on the cathode side is presented in Figure 4.74 and Figure 4.75.

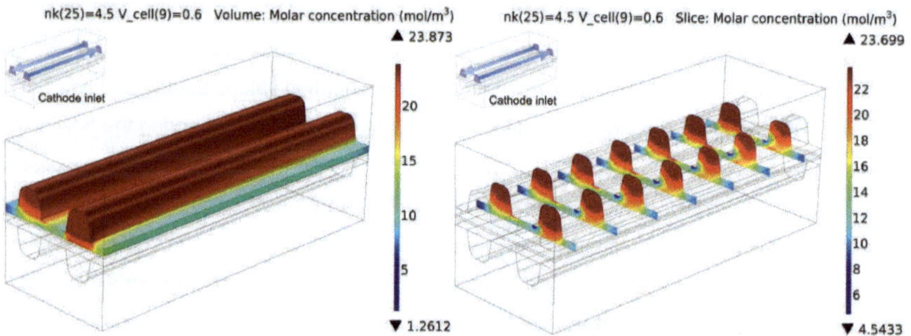

Figure 4.74 Molar concentration [mol/m³] of the nitrogen on the cathode

Figure 4.75 Molar concentration [mol/m³] of the nitrogen on the cathode in slices

The distribution of the nitrogen molar concentration is developed due to the mass conservation regarding the consumption, production and transport of oxygen and water correspondingly. Molar concentrations of the species on the cathode are induced by the electrochemical modeling and coupled variables of the physical interfaces as in the anode.

Density values on the anode and cathode side of the fuel cell are calculated utilizing the molar concentrations of the species. The computed density values on the anode and cathode are depicted in Figure 4.76 and Figure 4.77 respectively. The calculated density values are integrated into the momentum transport.

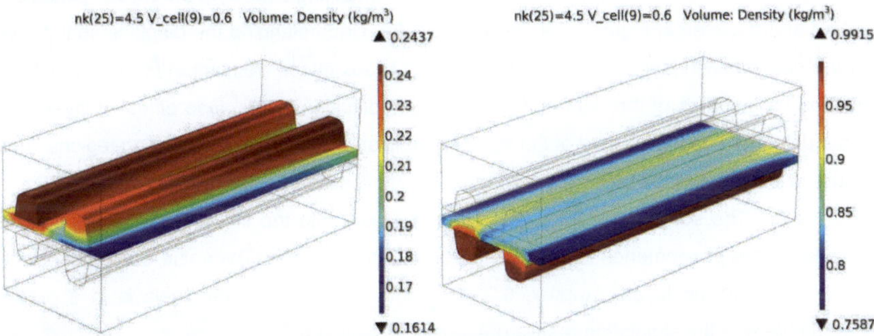

Figure 4.76 Density [kg/m³] on the anode   **Figure 4.77** Density [kg/m³] on the cathode

Von Mises stress distribution on a cut plane in the middle of the cross-section model can be seen in Figure 4.78. Compression pressure acts on the upper surface of the bipolar plate as boundary condition.

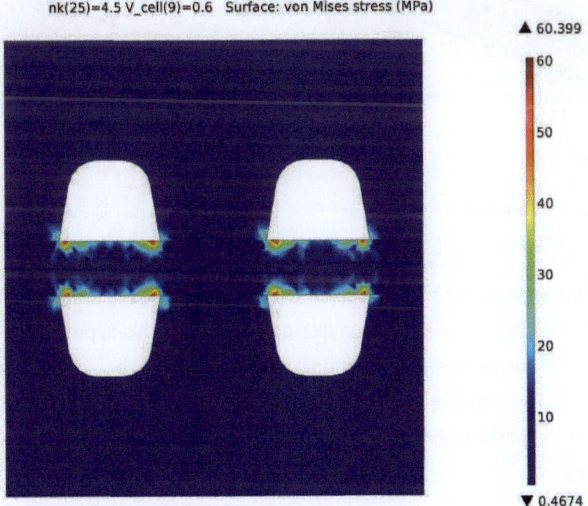

Figure 4.78 Von Mises Stress [MPa] on the cross-section model

Mechanical pressure can only be transferred through the bipolar plate lands on the gas diffusion layers. This results in a higher stress profile on the contacting edges of the bipolar plates to the gas diffusion layers as it can be realized in Figure 4.78. The geometry of the bipolar plate and acting compression force can be adapted in order to change the mechanical stress distribution on gas diffusion layers. The size of the flow field channel and bipolar plate lands can be modified in order to optimize the cell performance via changing the permeability and electrical resistance by preventing undesired deformations of the gas diffusion layer.

Displacement profiles of the gas diffusion layers at contacting surface to the bipolar plates are scaled and depicted in Figure 4.79 and Figure 4.80. Understanding the displacement profile on the gas diffusion layers together with the Von Mises stress distribution (see Figure 4.78) enables the analysis of the deformation of the gas diffusion layers. Gas diffusion layers are under different mechanical load of the flow channels and bipolar plate contacting regions. This results in a change of the deformation of the gas diffusion layers as realized in Figure 4.79 and Figure 4.80. The squeezing of the gas diffusion layers into the flow field channels can be remarked. This phenomenon is dealt in section 4.2 in stack size. The value of the squeezing of the gas diffusion layers (approx. 0.75[µm]) into the flow field channels is in a proper accordance with the simulation results presented in section 4.2.

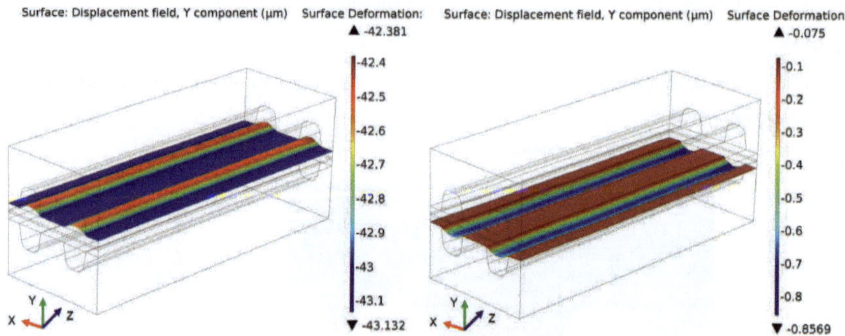

Figure 4.79 Total displacement [µm] of the gas diffusion layer surface (anode)

Figure 4.80 Total displacement [µm] of the gas diffusion layer surface (cathode)

Permeability distribution of the gas diffusion layers depicted in Figure 4.63 can be analyzed superiorly with the help of the stress and displacement distributions presented in Figure 4.78 and Figure 4.79-80 respectively. The squeezing of the GDL under bipolar plate channel and lands and its effect on the permeability can be observed together. The dependency of the GDL permeability on temperature and humidity can also be investigated, which is not considered within this study.

The simulation results for the analysis of the mass and momentum transports are presented for an equivalent compression pressure value of 4.5[MPa] and voltage value of 0.6[V]. The permeability of the gas diffusion layers and the mechanical stress distribution on the cross-section model are figured out supplementally. In order to evaluate the performance of the fuel cells, the polarization curve is utilized as a standard tool (see section 2.2.4 and 3.3). The simulation results are also to be compared with the measured data obtained from a fuel cell stack with the same size and components. As explained before the boundary conditions for the mass and momentum transports are provided by the measured data in order to perform the verification of the simulations. Material properties in the cross-section model are also defined same as the used cell components in the measured fuel cell stack. This fulfills the cross-section modeling and enables the verification with the help of measured data generated from a fuel cell stack with same properties.

The polarization curve is simulated by computing the average current density on the membrane under varying potential on the cathode as explained before. The measured polarization curve is also obtained in same manner (see section3.3) for a constant medium supply (with an average value of 4.01[l/min] hydrogen and 9.02[l/min] air respectively), which enables the comparison with the simulation results. The simulated polarization curve performed by a parametric study is plotted in Figure 4.81 together with the measured data.

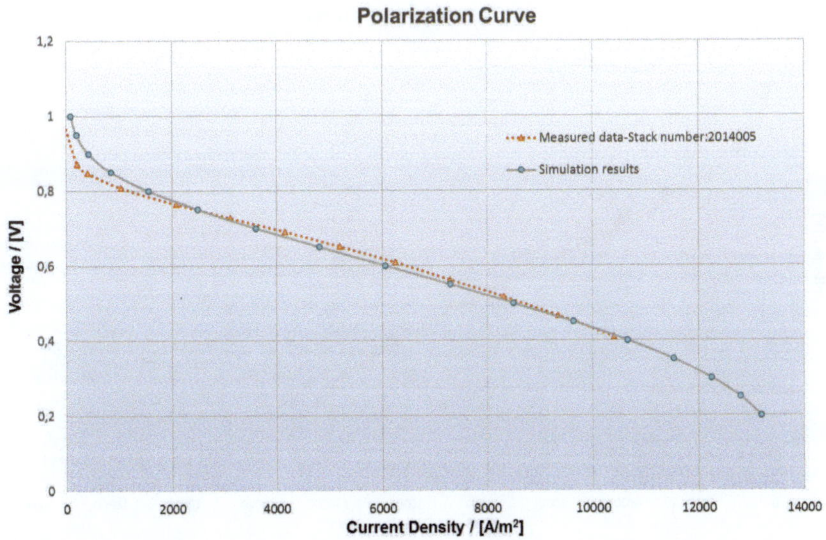

Figure 4.81 Simulated polarization curve in comparison with measured data

As it can be remarked in Figure 4.81, the simulation results show a proper accordance with measured data regarding several assumptions in the simulations including membrane proton conductivity and one phase transport as given in section 4.3.2.2. The deviance in the activation losses can be noticed between the simulated and measured polarization curves, which can be attributed to the parameters of electrochemical modeling for example effective binary diffusivities and exchange current density. The measurements have also several imprecisions for example sensor and mass flow inaccuracies and laboratory conditions like temperature. Temperature and humidity dependencies of material properties e.g. permeability of gas diffusion layers and two phase transport can be implemented into the model with the help of supplemental measurements for a further analysis.

The cross-section model is built precisely regarding the objective of this study. Mechanical pressure is parameterized in order to analyze its effect on the fuel cell performance as mentioned before. Generating the polarization curve for each predefined mechanical pressure provides the analysis of the fuel cell performance from mechanical point of view. The simulated polarization curves for a wide range of mechanical pressure values are plotted in Figure 4.82. The simulation interface enables the extension of the analysis for immeasurable cases, which contributes to the accurate analysis of the investigated variable. A polarization curve for an extreme compression pressure value of 8.1[MPa] or 0.01[MPa] is almost impossible to generate with measurements. However mathematical modeling enables this by a sensitivity analysis.

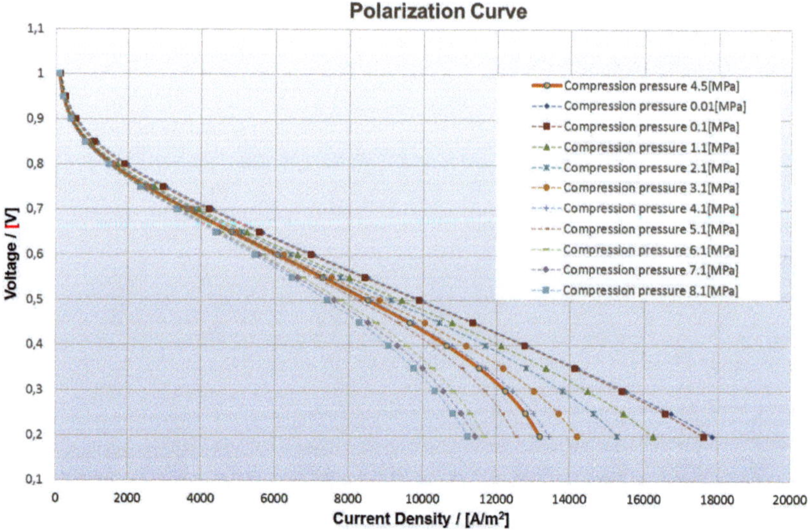

Figure 4.82 Simulated polarization curves for different compression pressure values

The polarization curve given in a bulk orange line is presented in Figure 4.82 for the equivalent compression pressure values as in the stack. A decline in the polarization curve denotes the performance loss. In contrary the incline of the polarization curve designates a gain in the performance. As it can be noticed in Figure 4.82 increasing the compression pressure results in performance loss for the fuel cell. This occurs due to the disturbance of the permeability of the gas diffusion layers. The decrease of the contact resistance due to the compression pressure increment doesn't affect the performance positively in comparison with the loss in the permeability. The modeling of charge transports can be improved by expanding model geometry and with the help of supplemental measurements for the material properties depending on temperature.

As it can be remarked in Figure 4.82, reaching the extremes for the mechanical compression values marginalizes its effect on the performance. The effect of the mechanical compression on the cell performance is increasing in higher current density regions due to the increased requirement for medium flow. Lower compression pressure values result in improved permeability values, which decreases the concentration loss in higher current density regions. The simulation results are in a proper accordance with selected literature [52-54, 120, 167, 176, 181].

Fuel cell stack design can be improved with the help the evaluation of the simulation results. Reducing the compression pressure values results in an improved cell performance as realized by the performed simulations. However the functionality of the fuel cell stack and the components are to be taken into consideration by stack design. The compression pressure has to be preserved by a value ensuring the tightness of the stack under fuel cell operation conditions. Thermal expansion of the fuel cell stack is taking place as investigated in section 3.1 and 4.2 with the help of measurements and performed simulations. Thermal expansion results in additional deformation of the sealings about 32%, which can change the tightness of the fuel cell stack. The tightness of the fuel cell stack depends not only on the sealing material and design also on the stack and bipolar plate design. The simulations provide supplemental information for the sealing design but can't ensure the tightness of the real fuel cell stack. This is to be considered by the fuel cell stack design. In order to reduce the required compression pressure adhesive bonding or welding of bipolar plates can be considered. The contact between bipolar plates and gas diffusion layers can be improved with the help of surface treatment, which could result in decrease of clamping force. Tightening with sealing materials applied outside of the fuel cells can also contribute to the reduced clamping force. Additional mechanical components, which yields even distributed compression pressure for the fuel cells can also be considered. This leads to improved sealing properties and reduced contact resistance by assuring a proper gas permeability for gas diffusion layers.

Consequently, two presented simulation models in this section enable together the explicit analysis of the fuel cell stack from mechanical point of view within the objective of this study. The stack-size simulation model with 5 fuel cells dealt in section 4.2 ensures the thermomechanical analysis of the fuel cell stack. Together with the stack model, the derived cross-section model (presented in section 4.3) is dealing with the electrochemical properties of fuel cells considering the mechanical compression. The performed simulations are verified with the measured data generated from a fuel stack with the same properties as presented in section 3.1. The interactions between the simulation models presented in this section and the use of the measurements for the verification and modeling purposes are illustrated in Figure 4.83 for a better understanding. This has ensured the utilization of the whole computational capacity for the analysis of mechanical properties of the fuel cell stacks.

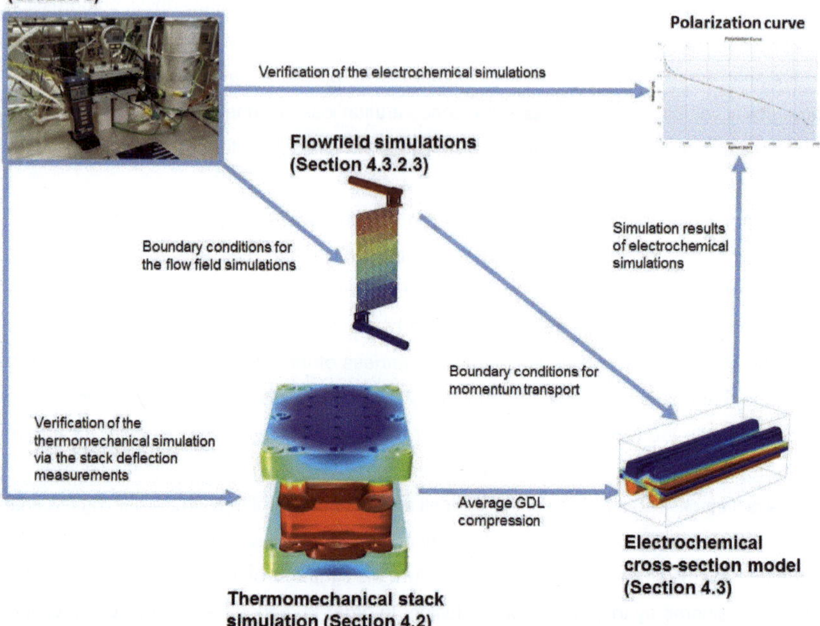

Figure 4.83 Interactions between simulation and measurements

The computational models can be developed further by improving the material properties and electrochemical modeling. Implementation of the thermal modeling together with the humidity and membrane swelling effect can deliver better results.

The simulation models can be utilized for other PEM fuel cell stacks by following the same guideline performed in this study. The models can be applied for the development of existing stacks but also for the stack designs within prototyping process. The guideline followed in this study for the mechanical analysis of fuel cell stacks can also be utilized for the similar applications such as electrolysis and flow batteries.

5 Conclusions and Outlook

A systematic investigation of the PEM fuel cell stack design from a mechanical point of view was performed in this study with the help of measurements and simulations. Analysis of the mechanical characteristics of the fuel cell stack contributes to the development of a design procedure for fuel cell stack design.

A fuel cell stack is to be held mechanically together to connect the consisting cells in serial. In addition to the electrical circuit configuration, mechanical clamping ensures the tightness and reduces electrical resistance. Assuring the tightness of the stack under the operating conditions is a key task to be achieved. Over-compressing of the fuel cell components disturbs their functionalities such as the permeability of the gas diffusion layers. Over-compressing the fuel cell stack can also result in damage of the stack components. An even stress distribution in a fuel cell stack is to be achieved in order to assure same properties for the whole active area of each fuel cell for an optimal performance. Desired compression pressure values are to be accomplished for the whole active area of the fuel cell. Mechanical properties of the fuel cell stack are to be examined properly in order to understand these key mechanical requirements. Understanding the fuel cell stack design properties from a mechanical point of view leads to the optimization process for the examined stack design and components.

A standard fuel cell stack design of the fuel cell research center ZBT with an active area of 50 cm^2 was analyzed within this work. A stack with 5 cells was focused on in this work in order to assure the observation of stack characteristics and utilization of experience in fuel cell stacks with same number of cells. The mechanical design principle of the ZBT standard fuel cell stack is based on compression of the cells with the help of endplates and tie rods.

Internal stress distribution within the fuel cell stack was measured with the help of pressure measuring films under non-operating conditions. Within this study an experimental set-up was established to be able to analyze the mechanical characteristics of the fuel cell stack under operating conditions. The deflection of the fuel cell stack due to the thermal expansion was measured with the help of the experimental set-up under operating conditions. The effect of cell operating parameters on the mechanical properties of the fuel cell stack was investigated supplementally. Cyclic measurements were additionally performed in order to investigate the fuel cell stack mechanics under dynamic load. Shrinkage of the fuel cell stack was observed due to the plastic deformation occurring in the fuel cell stack by the thermal expansion and compression.

The deflection of the fuel cell stack and its dependency on the operating parameters have been introduced with the help of performed measurements. The deflection measurements could be performed for a single reference point, which restricts the analysis of the mechanical characteristics. Simulation interface enables the evaluation of the immeasurable cases, which deliver supplementary information for the analysis of the fuel cells. In order to extend the mechanical analysis of the fuel cells, two computational models were established. A large-scale fuel cell stack model was built for the thermomechanical analysis. The second model was built on a cross-section geometry for the electrochemical analysis. Detailed literature survey of the existing fuel cell models in terms of the mechanical analysis were performed by modeling efforts within the framework of this study.

In order to examine the deflection of the fuel cell stack, a large-scale thermomechanical model was established. 3D geometry of the exemplary stack with 5 cells was utilized by generating the simulation model. The 3D model contains all geometrical features excluding small fillets and stack components, which are irrelevant from the mechanical point of view. The simulation model was built with regard to the geometrical properties and problem definition. The computational capacity was also considered by performed simulations.

Structural and thermomechanical characteristics of the fuel cell stack were investigated with the help of the developed 3D model. The simulation results were verified by utilizing the measured data provided by the experimental set-up. The internal stress distribution of the fuel cell stack was examined. The comparison of the structural and thermomechanical simulations corresponds to a proper understanding of the effect of thermal expansion on the fuel cell stack. Expanding of the stack components depends not only on the temperature profile also on the thermal expansion coefficient and geometries of the components. Thermal expansion uniforms the internal stress distribution of the fuel cell stack but increase the mechanical loading on the fuel cell components.

Buckling of the fuel cell stack was observed with the help of the simulations. The analysis of the mechanical load acting on the stack components was evaluated by the simulations. The effect of the stack design and geometrical properties of stack Components e.g. bipolar plates and endplates on the mechanical characteristics of the fuel cell stack was properly realized. An optimization procedure for the fuel cell stack can also be performed regarding the topology and dimensioning of the components such as the endplates.

Simulating the effect of the mechanical compression on the cell performance could not be performed in stack size due to the restrictions of the computational capacity. In order to analyze the cell performance depending on the compression pressure, a second model was built for a cross-section geometry of a single fuel cell. Compression pressure was implemented into the developed electrochemical model for the mechanical analysis. Boundary settings and the material modeling were precisely performed. Flow fields for the anode and cathode side were simulated from a fluid dynamics point of view in order to define the boundary settings of the cross-section model. The interaction of the physical interfaces was built with the help of the boundary and domain settings together with the material modeling. The developed multiphysical cross-section model was simulated under varying cathodic potentials for diverse compression pressure values. The cell performance by varying compression pressure values were available with the help of these simulations. The polarization curves generated by the electrochemical simulations were investigated in detail for a better understanding of the influence of mechanical compression on the cell performance and properties of cell components. The simulated polarization curve was verified with the help of the measured data attained by the exemplary fuel cell stack with the same size and properties.

Several assumptions have been made by the performed simulations, which must have been taken into consideration by the evaluation of the simulation results. The two phase transformation was neglected in the electrochemical modeling and the water was assumed to be in gas phase only. Membrane was assumed to be fully humidified by the electrochemical simulations.

It was realized by the simulations that increasing the compression pressure value results in a performance loss due to the disturbance of the gas diffusion layer permeability. Compression of the fuel cell reduces the contact resistance values, which results in a performance increase. This cannot compensate the disturbance in the gas permeability by increasing the compression pressure values.

As a result, the fuel cell stack is to be compressed with a clamping force which doesn't exceed the appropriate values for the gas diffusion layer permeability. The tightness of the fuel cell stack is to be assured under stack operating conditions.

In this study, the 3-dimensional models were built containing the geometrical features in detail, which was achieved first time. This enables the exact analysis and verifications of the simulation results with measured data therewith the evaluation of the models. The thermal expansion of a fuel cell stack was measured first time under real operating conditions. The effect of the cell operating conditions on the mechanical properties of the fuel cell stack was also investigated with the help of the measurements. The first thermomechanical PEM fuel cell stack model is established in detail. The first 3-dimensional electrochemical model containing two adjacent channels was realized from a mechanical point of view. This enables the analysis of the influence of the mechanical compression on the cell performance.

The models were established for further utilization purposes. The models can also be extended for the electrochemical analysis via heat transfer including the membrane humidification and enhanced by an improved material modeling.

The analysis contributes to the understanding of fuel cell mechanisms from a mechanical point of view and thus leads to an increase in fuel cell performance. In addition to that getting an optimal and even distributed compression pressure value for stack components under dynamic load extends the durability of the fuel cells. This eliminates a significant obstacle on the roadway of the market launch of the fuel cells as products.

The analysis of an existing exemplary fuel cell stack with the help of experiments and computational models denotes not only the optimization of components also the development of a design procedure. Any fuel cell stack concept and similar applications such as electrolysis can also be properly investigated during the design process with the same analysis steps performed in this study. Design procedure starts with the analysis of the stack with the help of thermomechanical computations and follows with the dimensioning and design of the stack components. This is to be taken into account by the fuel cell stack design. Acquiring the compression pressure distribution within the fuel cell stack leads to the performance analysis with the help of the extended electrochemical simulations. The analysis of the stack performance yields modification of the stack and component design and closes the design procedure.

As a result the mechanical pressure distribution profile and its effect on fuel cell performance were examined and the mechanical analysis of the standard PEM fuel cell stack design of ZBT was performed with the help of experiments and computational methods. The developed design procedure for the standard 50 cm² fuel cell stack of ZBT can be used for any size of fuel cell stack design and other similar applications. The developed design procedure in this study has been utilized for different fuel cell stack designs during the prototyping process.

6 References

1. Stern, N., The Economics of Climate Change, Cambridge University Press, 2008

2. http://www.opec.org/opec_web/static_files_project/media/downloads/publications/WOO2012.pdf

3. Management of fluctuations in wind power and CHP comparing two possible Danish strategies, Lund, H., Clark, W. W., Energy, 27 (5) pp. 471–483, 2002

4. Lund, H., Kempton, W., Integration of renewable energy into the transport and electricity sectors through V2G, Energy Policy, 36, Pages 3578–3587, 2008

5. Mathiesen B. V., Lund, H., Comparative analyses of seven technologies to facilitate the integration of fluctuating, , IET Renew. Power Gener., Vol. 3, pp. 190–204, 2009

6. Jensen, S. H., Larsen, P. H., Mogensen, M., Hydrogen and synthetic fuel production from renewable energy sources, Int. J. Hydrogen Energy, pp. 3253–3257, 2005

7. Sherif, S. A., Barbir, F., Veziroglu, T. N., Wind energy and the hydrogen economy-review of the technology, Sol. Energy, 78, pp. 647–660, 2005

8. Chu, S., Majumdar, A., Opportunities and challenges for a sustainable energy future, Nature, 488, 294-303, 2012

9. Heinzel, A., Brennstoffzellen: Entwicklung, Technologie, Anwendung, Heidelberg: Müller, (2006)

10. Barbir, F., PEM Fuel Cells: Theory and Practice, Amsterdam: Elsevier, (2005)

11. Wang, H. et al., A Review of PEM Fuel Cell Durability: Degradation Mechanisms and Mitigation Strategies, J. Power Sources, 184, 104-119, (2008)

12. Wang, H. et al. A Review of Polymer Electrolyte Membrane Fuel Cell Durability Test Protocols, J. Power Sources, 196, 9107-9116, (2011)

13. Sammes, N., Fuel Cell Technology: Reaching Towards Commercialization, London: Springer, (2006)

14. Yan, W. M., Wang, X. D., Mei, S. S., Peng, X. F., Guo, Y. F., Su, A., Effects of operating temperatures on performance and pressure drops for a 256 cm2 proton exchange membrane fuel cell: An experimental study, J. Power Sources, 185, 1040-1048,(2008)

15. Wu, H. W., Gu, H. W., Analysis of operating parameters considering flow orientation for the performance of a proton exchange membrane fuel cell using the Taguchi method, J. Power Sources, 195, 3621-3630, (2010)

16. Jo, Y. Y., Cho, EA., Kim, J. H., Lim, T. H., Oh, I., Kim, S., Kim, H., Jang, J. H., Degradation of polymer electrolyte membrane fuel cells repetitively exposed to reverse current condition under different temperature, J. Power Sources, 196, 9906-9915, (2011)

17. Yan, Q., Toghiani, H., Causey, H., Steady State and Dynamic Performance of Proton Exchange Membrane Fuel Cells (PEMFCs) under Various Operating Conditions and Load Changes, J. Power Sources, 161, 492-502, (2006)

18. O'Hayre, R. P., Fuel Cell Fundamentals, Hoboken: Wiley, (2006)

19. Mench, M. M., Fuel Cell Engines, Hoboken: Wiley, (2008)

20. Wang, Y., Chen, K. S., Mishler, J., Cho, S. C., Adroher, X. C., A Review of Polymer Electrolyte Membrane Fuel Cells,: Technology, Applications, and Needs on Fundamental Research, Appl. Energ., 88, 981-1007, (2011)

21. Hoogers, G., Fuel Cell Technology Handbook, Boca Raton: CRC Press, (2003)

22. Tribecraft AG, Endplate for a Stack of Fuel Cells, WO2004075350A1, 2004

23. Lee, J. K., Cheong, S. I., Clamping Apparatus of a Fuel Cell Stack Capable of Applying Uniform Pressure Through Total to The Entire Area of a Seperator Constituting a Fuel Cell Stack, KR102010020715A, 2010

24. Siemens AG, Flexibles Zwischenelement für einen Brennnstoffzellenstack, DE10003528A1, 2001

25. Elring Klinger AG, Fuel Cell Stack, US20080070068A1, 2008

26. GM Global Technology Operations, Inc., Fuel Cell Stack Compression Enclosure Apparatus, US20110244355A1, 2011

27. Hyundai Motor Company, Fuel Cell Stack Assembly, US20090226793A1, 2009

28. Hyundai Motor Company, Fuel Cell Stack Clamping Device, US000007776489B2, 2010

29. Kyu, T.C et al., Fastening Mechanism of a Fuel Cell Stack, US20050064268A1, 2005

30. Ballard Power Systems Inc., Electrochemical Fuel Cell Stack with an Improved Compression Assembly, US6190793B1, 2001

31. Ballard Power Systems Inc., Electrochemical Fuel Cell Stack with Compression Bands, WO001998022990A1, 1998

32. Fuel Cell Energy Inc., Fuel Cell Stack Compression System, WO002002019456A1, 2002

33. Toyota-cho, Fuel Cell Stack, WO002005053080A1, 2005

34. Berning, T., Djilali, N., Three-dimensional Computational Analysis of Transport Phenomena in a PEM Fuel Cell – A Parametric Study, Journal of Power Sources, 124, 440-452, (2003)

35. Kandlikar S. G., Lu, Z., Thermal management issues in a PEMFC stack – A brief review of current status, Appl. Therm. Eng., 29, 1276–1280, (2009)

36. Zhang, G., Kandlikar, S.G., A critical review of cooling techniques in proton exchange membrane fuel cell stacks, Int. J. Hydrogen Energy, 37, 2412-2429, (2012)

37. Firat, E., Bandlamudi, G., Beckhaus, P., Heinzel, A., Heat Pipe Assisted Thermal Management of a HT PEMFC Stack, Proceedings of European COMSOL conference, Milan, Italy, (2012)

38. Wu, J., Yuan, X. Z., Martin, J. J., Wang, H., Zhang, J., Shen, J., Wu, S., Merida, W., A review of PEM fuel cell durability: Degradation mechanisms and mitigation strategies, J. Power Sources, 184, 104-119,(2008)

39. Heinzel, A., Mahlendorf, F., Jansen, C., Bipolar Plates, Encyclopedia of Electrochemical Power Sources, 8010-816, (2009)

40. Firat, E., Kühnemann, L., Siegel, C., Beckhaus, P., Heinzel, A., Finite Element Approach for Optimizing the Cooling of Metallic Bipolar Plates for Fuel Cell Applications, Proceedings of European COMSOL conference, Paris, France, (2010)

41. Wind, J., Späh, R., Kaiser, W., Böhm, G., Metallic Bipolar Plates for PEM Fuel Cells, Journal of Power Source, 105, 256-260, (2002)

42. Myung, S.-T., Sakurada, S., Kumagai, M., Yashiro, H., A Promising Alternative to PEMFC Graphite Bipolar Plates: Surface Modified Type 304 Stainless Steel with TiN Nanoparticles and Elastic Styrene Butadiene Rubber Particles, Fuel Cells 10, 4, 545-555, (2010)

43. Tawfik, H., Hung, Y., Mahajan, D., Metal bipolar plates for PEM fuel cell—A review, J. Power Sources, 163, 755-767,(2007)

44. Yuan, X. Z., Wang, H., Zhang, J., Wilkinson, D. P., Bipolar Plates for PEM Fuel Cells - From Materials to Processing, J. New. Mat. Electrochem. Systems, 8, 257-267 (2005)

45. Middleman, E., Kout, W., Vogelaar, B., Lenssen, J., De Waal, E., Bipolar Plates for PEM Fuel Cells, J. Power Sources, 118, 44-46 (2003)

46. Müller, A., Kauranen, P., Von Ganski, A., Hell, B., Injection Moulding of Graphite Composite Bipolar Plates, J. Power Sources, 154, 467-471 (2006)

47. Hermann, A., Chaudhuri, T., Spagnol, P., Bipolar Plates for PEM Fuel Cells: A review, Int. J. Hydrogen Energy, 30, 1297-1302, (2005)

48. Li, X., Sabir, I., Review of Bipolar Plates in PEM Fuel Cells: Flow-Field Designs, Int. J. Hydrogen Energy, 30, 359-371 (2005)

49. Mehta, V., Cooper, J. S., Review and Analysis of PEM Fuel Cell Design and Manufacturing, J. Power Sources, 114, 32-53, (2003)

50. Kreuz, C., PEM-Brennstoffzellen mit Spritzgegossenen Bipolarplatten aus hochgefülltem Graphit-Compound, Universität Duisburg-Essen, Dissertation, (2008)

51. Heinzel, A., Mahlendorf, F., Niemzig, O., Kreuz, C., Injection Moulded Low Cost Bipolar Plates for PEM Fuel Cells, J. Power Sources, 131, 35-40, (2004)

52. Xing, X. Q., Lum, K. W., Poh, H.J., Wu, Y. L., Optimization of assembly clamping pressure on performance of proton-exchange membrane fuel cells, J. Power Sources, 195, 62-68 (2010)

53. Zhou, Y., Lin, G., Shih, A.J., Hu, S. J., Multiphysics Modeling of Assembly Pressure Effects on Proton Exchange Membrane Fuel Cell Performance, J. Fuel Cell Sci. Tech., 6(4), 041005, (2009)

54. Brett, D. J. L., Shearing, P. R., Millichamp, J., Mason, T. J., A Study of The Effect of Compression on The Performance of Polymer Electrolyte Fuel Cells Using Electrochemical Impedance Spectroscopy and Dimensional Change Analysis, Int. J. Hydrogen Energy, 38, 7414-1422, (2013)

55. Liu, H., Zhang, G., Guo, L., Ma, L., Simultaneous Measurement of Current and Temperature Distribution in a Proton Exchange Membrane Fuel Cell, J. Power Sources, 195, 3597-3604, (2010)

56. Li, X., Karimi, G., Alaefour, I. E., Jiao, K., Simultaneous Measurement of Current and Temperature Distributions in a Proton Exchange Membrane Fuel Cell During Cold Start Process, Electrochim Acta, 56, 2967-2982, (2011)

57. Tao, W. He, Y., Chen, L., Cao, T., Lin, H., In Situ Measurement of Temperature Distribution within a Single Polymer Electrolyte Membrane Fuel Cell, Int. J. Hydrogen Energy, 37, 11871-11886, (2012)

58. Liu, W., Wan, Z., Tu, Z., Yu, Y., Zhang, H., Liu, Z., Pei, H., In Situ Measurement of Temperature Distribution in Proton Exchange Membrane Fuel Cell I a Hydrogen-Air Stack, J. Power Sources, 227, 72-79, (2013)

59. Brett, D. J. L., Shearing, P. R., Reisch, T., Fraga, E. S., Obeisun, O. A., Meyer, Q., Robinson, J. B., Noorkami, M., Effect of Temperature Uncertainity on Polymer Electrolyte Fuel Cell Performance, , Int. J. Hydrogen Energy, 39, 1439-1448, (2014)

60. Vahidi, A., Schmittinger, W., A Review of The Main Parameters Influencing Long-term Performance and Durability of PEM Fuel Cells, J. Power Sources, 180, 1-14, (2008)

61. Ji, M., Wei, Z., A Review of Water Management in Polymer Electrolyte Membrane Fuel Cells, Energies, Volume 2, (2009)

62. Debenjak, A. et al., Detection of Flooding and Drying inside a PEM Fuel Cell Stack, Stroj Vestn-J Mech E, 59, (2013)

63. Hanna, S., Elliott, J. A., Elliott, A. M. S., Cooley, G. E., The Swelling Behaviour of Perfluorinated Ionomer Membranes in Ethanol/Water Mixtures, Polymer, 42, 2251-2253, (2001)

64. Dammak, L. et al., Swelling and Permeability of Nafion 117 in Water-Methanol Solutions: An Experimental and Modeling Investigation, J. Membrane Sci., 377, 54-64, (2011)

65.   E. I. du Pont de Neumours and Company

66.   Fujifilm Corporation

67.   Regalla, S. P., Computer Aided Analysis and Design, I. K. Int. Publishing House Pvt. Ltd., (2010)

68.   Matlab/Simulink Software, see <http://www.mathworks.com>

69.   Dymola Software, see <http://www.dymola.com>

70.   Modelica Software, see <http://www.modelica.com>

71.   Scilab Software, see'<http://www.scilab.org>

72.   Pukrushpan, J., T., Stefanopoulou, A.G., Peng, H., Control of Fuel Cell Power Systems, 2nd ed., Springer, (2005)

73.   Spiegel, C., PEM Fuel Cell Modeling and Simulation Using Matlab, Elsevier, (2008)

74.   Gou, B., Na, W. K., Diong, B., Fuel Cells: Modeling, Control, and Applications, CRC Press, (2010)

75.   Nehrir, M. H., Wang, C., Modeling and Control of Fuel Cells, John Wiley & Sons Inc. (2009)

76.   Srinivasan, S., Fuel Cells: From Fundamentals to Applications, Springer, (2006)

77.   Graf, C., Vath, A., Nicoloso, N., Modeling of The Heat Transfer in a Portable PEFC System within MATLAB-Simulink, J. Power Sources, 155, 52-59, (2003)

78.   Ceraolo, M., Miulli, C., Pozio, A., Modelling Static and Dynamic Behaviour of Proton Exchange Membrane Fuel Cells on The Basis of Electro-Chemical Description, J. Power Sources, 113, 131-144, (2003)

79.   Yip (ed.), S., Handbook of Materials Modeling, Volume:1: Methods and Models,1-32, Netherlands: Springer, (2005)

80.   Grossmann, C., Ross, H. G., Stynes, M., Numerical Treatment of Partial Differential Equations, Berlin: Springer-Verlag, (2005)

81.   Schäfer, M., Computational Engineering: Introduction to Numerical Methods, Berlin: Springer, (2006)

82.   Pham, D., Comparison of Finite Volume and Finite Difference Methods and Convergence Results for Finite Volume Shemes, Indiana University (USA), Dissertation, (2007)

83.   Fasshauer, G. E., Meshfree Approximation Methods with MATLAB, Singapore: World Scientific Publishing Co. Pte. Ltd., (2007)

84.   Liu, G. R., Mesh Free Methods: Moving Beyond the Finite Element Method, Boca Raton: CRC Press., (2003)

85.   Pepper, D. W., Heinrich, J. C., The Finite Element Method-Basic Concepts and Applications, United States of America: Hemisphere Publishing Corporation, (1992)

86.   Munz, C. D., Bolemann, T., Staudenmaier, M., Numerik partieller Differentialgleichungen, Universität Stutgart, (2012)

87.   Goering, H., Roos, H. G., Tobiska, L., Die Finite-Elemente-Methode für Anfänger, Magdeburg/Dresden: Wiley-VCH Verlag GmbH & Co. KGaA, (2009)

88.   Schröder, C., Fromme, L., COMSOL Multiphysics-So funktioniert FEM!, Göttingen: COMSOL, (2012)

89.   Jenni, T., Einführung in die Finite-Elemente-Methode, Zürich, (2011)

90.   Nikishkov, G. P., Introduction to The Finite Element Method, Lecture Notes: University of Aizu, (2007)

91. Andersson, B. et al., Computational Fluid Dynamics For Engineers, Cambridge: Cambridge University Press, (2012)

92. Wendt, J. F. et al., Computational Fluid Dynamics: An Introduction, Heidelberg: Springer, (2009)

93. Löhner, R., Applied CFD Techniques: An Introduction based on Finite Element Methods, Chichester: Wiley, (2008)

94. Klein, B., Grundlagen und Anwendungen der Finite-Element-Methode im Maschinen-und Fahrzeugbau, Wiesbaden: Vieweg+Teubner Verlag, (2012)

95. Zienkiewicz, O. C., Taylor, R., Zhu, J., The Finite Element Method. 1. Its Basis and Fundamentals, Amsterdam: Elsevier, (2008)

96. Zienkiewicz, O. C., Taylor, R., Zhu, J., The Finite Element Method. 2.The Finite Element Method for Solid and Structural Mechanics, Amsterdam: Elsevier, (2006)

97. Zienkiewicz, O. C., Taylor, R., Zhu, J., The Finite Element Method. 3. For Fluid Dynamics, Amsterdam: Elsevier, (2006)

98. Steinke, P., Finite-Elemente-Methode: Rechnergestützte Einführung, Heidelberg: Springer, (2010)

99. Knothe, K., Wessels, H., Finite Elemente: Eine Einführung für Ingenieure, Berlin: Springer, (2008)

100. Tong, P., Rossettos, J. N., Finite Element Method: Basic Technique and Implementation, Mineola: Dover Publ., (2008)

101. Merkel, M., Öchsner, A., Eindimensionale Finite Elemente: Ein Einstieg in die Methode, Heidelberg: Springer, (2010)

102. Rieg, F., Hackenschmidt, R., Alber-Laukant, B., Finite-Elemente-Analyse für Ingenieuere, München: Hanser, (2012)

103. Smith, I. M., Griffiths, D. V., Programming the Finite Element Method, Chiecester,: Wiley, (2004)

104. Gockenbach, M. S., Understanding and Implementing the Finite Element Method, Philadelphia: SIAM, (2006)

105. Reddy, J. N., An Introduction to the Finite Element Method, New York: McGraw-Hill, (1984)

106. Reddy, J. N., Gartling, D. K., The Finite Element Method in Heat Transfer and Fluid Dynamics, Boca Raton: CRC Press, (2010)

107. Dhondt, G., The Finite Element Method for three-Dimensional Thermomechanical Applications, Chichester: Wiley, (2004)

108. Thompson, E. G., An Introduction to the Finite Element Method: Theory, Programming, and Applications, Hoboken,NJ: Wiley, (2005)

109. Rao, S. S., The Finite Element Method in Engineering, Amsterdam: Elsevier/Butterworth Heinemann, (2005)

110. COMSOL Multiphysics Modeling Guide Ver. 3.5a (2008)

111. Zimmermann, W. B. J., Multiphysics Modelling with Finite Element Methods, World Scientific Publishing Company Inc., (2006)

112. Pryor, R. W., Multiphysics Modeling Using COMSOL: A First Principles Approach, USA: Jones and Bartlett Publishers, LLC., (2011)

113. Martemianov, S., Bograchev, D., Gueguen, M., Grandidier, J. C., Stress and Plastic Deformation of MEA in Fuel Cells: Stress Generated During Cell Assembly, J. Power Sources, 180, 393-401, (2008)

114. Kusoglu, A., Santare, M. H., Karlsson, A. M., Cleghorn, S., Johnson, W. B., Numerical Investigation of Mechanical Durability in Polymer Electrolyte Membrane Fuel Cells, J. Electrochem. Soc., 157, 705-713, (2010)

115. Lee, S. J., Hsu, C. D., Huang, C. H., Analyses of The Fuel Cell Stack Assembly Pressure, J. Power Sources, 145, 353-361, (2005)

116. Mélé, P., Carral, C., A Numerical Analysis of PEMFC Stack Assembly Through 3D Finite Element Model, Int. J. Hydrogen Energy, 39, 4516-4530, (2014)

117. Park, S. et al., Simulation and Experimental Analysis of The Clamping Pressure Distribution in A PEM Fuel Cell Stack, Int. J. Hydrogen Energy, 38, 6481-6493, (2013)

118. Mikkola, M. S., Koski, P. A., Modeling the Internal Pressure Distribution of a Fuel Cell, Proceedings of the European COMSOL Conference, Milan, 2009

119. Firat, E., Beckhaus, P., Heinzel, A., Finite Element Approach for the Analysis of the Fuel Cell Internal Stress Distribution, Proceedings of European COMSOL conference, Stuttgart, Germany, (2011)

120. Liu, H. , Higier A., Ge, J., Effect of Gas Diffusion Layer Compression on PEM Fuel Cell Performance, J. Power Sources, 159, 922-927, (2006)

121. Montanini, R.,Squadrito, G., Giacoppo, G., XIX IMEKO World Congress Fundamental and Applied Metrology, September 6-11, (2009)

122. Wang, X., Song, Y., Zhang, B., Experimental Study on Clamping Pressure Distribution in PEM Fuel Cells, J. Power Sources, 179, 305-309, (20008)

123. Carral, C., Charvin, N., Trouvé, H,. Mélé, P., An Experimental Analysis of PEMFC Stack Assembly Using Strain Gage Sensors, Int. J. Hydrogen Energy, 39, 4493-4501, (2014)

124. Von Mark S. Gockenbach, Partial differential equations: analytical and numerical methods, (2002)

125. COMSOL Multiphysics® Structural Mechanics Module User's Guide, pg. 179 (2008)

126. Kleemann, J., Finsterwalder, F., Tillmetz, W., Characterisation of Mechanical Behavior and Coupled Electrical Properties of Polymer Electrolyte Membrane Fuel Cell Gas Diffusion Layers, J. Power Sources, 190, 92-102 (2009)

127. VDI-Richtlinie 2230; Systematische Berechnung hochbeanspruchter Schraubenverbindungen (Abschnitt 13.1.9)

128. Wittel, H. et al., Roloff/Matek: Maschinenelemente: Normung Berechnung Gestaltung, Springer, 21. Auflage, (2013)

129. COMSOL Multiphysics® Heat Transfer Module User's Guide, pg. 53, (2013)

130. Incropera, F. P., DeWitt, D. P., Bergmann, T. L., Lavine, A. S., Fundamentals of Heat and Mass Transfer, John Wiley & Sons, Sixth Edition, (2006)

131. Gibbins, J., Thermal Contact Resistance of Polymer Interfaces, University of Waterloo, Canada, Dissertation, (2006)

132. Holman, J. P., Heat Transfer, USA: McGraw_Hill, (1997)

133. Bahrami, M., Yovanovich, M., Marotta, E. E., Modeling of Thermal Joint Resistance of Polymer-Metal Rough Interfaces, Proceedings of IMECE, California, USA, (2004)

134. http://www.matweb.com

135. ZBT GmbH (Zentrum für Brennstoffzellentechnik in Duisburg)-Center for Fuel Cell Technology in Duisburg

136. DIN EN ISO 178, Kunststoffe-Bestimmung der Biegeeigenschaften (ISO 178:2010); Deutsche Fassung EN ISO 178:2010

137. Otmani, N., Morin, A., Di Iorio, S., Rouillon, L., Delette, G., Blachot, J.F., Gebel, G., Besse, S., Stresses in Nafion® Membranes During Fuel Cell Operation, Fundamentals and Developments of Fuel Cells Con., (2008)

138. Shah, A. A., Ralph, T. R., Walsh, F. C., Modeling and Simulation of the Degradation of Perfluorinated Ion-Exchange Membranes in PEM Fuel Cells, J.Electrochem. Soc., 156, (4) B465-B484 (2009)

139. Comsol Material Library COMSOL Multiphysics® (4.3.b)

140. Chao, Y. J., Tan, J., Li, X., Van Zee, J. W., Degradation of Silicone Rubber under Compression in a Simulated PEM Fuel Cell Environment, J. Power Sources, 172, 782-789, (2007)

141. Callister, J.D. W., Materials Science and Engineering an Introduction, John Wiley & Sons Inc., Page: 789-816, (2000)

142. Dynatec SA

143. http://www.azom.com

144. Wacker Chemie AG.

145. Toray Industries, Inc.

146. Hibbeler, R. C., Mechanics of Materials, New Jersey: Prentice Hall, Inc., (2000)

147. COMSOL Multiphysics User's Guide Ver. 3.5a (2008)

148. Montanini, R., Squadrito, G., Giacoppo, G., Assesment of Fuel Cell's Enplate Out of Plane Deformation Using Digital Image Correlation, in: Neri et al., Sensors and Microsystems, Springer Verlag, Dordrecht (NL), 443-447, (2011)

149. Saha, L. K., Tabe, Y., Oshima, N., Effect of GDL Deformation on The Pressure Drop of Polymer Electrolyte Fuel Cell Seperator Channel, J. Power Sources, 100-107, (2012)

150. Mélé, P., Carral, C., Charvin, N., Trouvé H., An Experimental Analysis of PEMFC Stack Assembly Using Strain Gage Sensors, Int. J. Hydrogen Energy, 4493-4501, (2014)

151. Siegel, C., Review of Computational Heat and Mass Transfer Modeling in Polymer-Electrolyte-Membrane (PEM) Fuel Cells, Energy, 1331-1352, (2008)

152. Biyikoglu, A., Review of Proton Exchange Membrane Fuel Cell Models, Int. J. Hydrogen Energy, 1181-1212, (2005)

153. Cheddie, D., Munroe, N., Review and Comparison of Approaches to Proton Exchange membrane Fuel Cell Modeling, J. Power Sources, 147, 72-84, (2005)

154. Secanell, M., Wishart, J., Dobson, P., Computational Design and Optimization of Fuel Cells and Fuel Cell Systems: A Review, J. Power Sources, 196, 3690-3704, (2011)

155. Wang, C. Y., Fundamental Models for Fuel Cell Engineering Chem. Rev., 104, 4727-4766, (2004)

156. Han, I. S., Lim, J., Jeong, J., Shin, H. K., Effect of Serpentine Flow-Field Designs on Performance of PEMFC Stacks for Micro-CHP Systems, Renew. Energ., 54, 180-188, (2013)

157. Meng, H., Song, G. H., Numerical Modeling and Simulation of PEM Fuel Cells: Progress and Perspective, Acta Mech Sinica, 29, 318-334, (2013)

158. Shimpalee, S., Beuscher, U., Van Zee, J. W., Investigation of Gas Diffusion Media Inside PEMFC Using CFD Modeling, J. Power Sources, 163, 480-489, (2006)

159. Wang, C. Y., Wang, Y., Ultra Large-Scale Simulation of Polymer Electrolyte Fuel Cells, J. Power Sources, 153, 130-135, (2006)

160. Wang, J., Simulation of PEM Fuel Cells by OpenFOAM, Heat and Mass Transport, MVK160, (2014)

161. Djilali, N., Nguyen, P. T., Berning, T., Computational Model of a PEM Fuel Cell with Serpentine Gas Flow Channels, J. Power Sources, 130, 149-157, (2004)

162. Fontes, E., Bryne, P., Schumacher, J., Vasileiades, N., Parhammar, O., Schmmitz, A., Influence of the Catalyst Layer on the Performance of the PEMFC Cathode, Technical Proceedings of the 2003 nanotechnology Conference and Trade Show, 3, 478-481, (2003)

163. Djilali, N., Computational Modelling of Polymer Electrolyte Membrane (PEM) Fuel Cells: Challenges and Opportunities, Energy, 32, 269-280, (2007)

164. Stenger, H. G., Güvelioglu, G.H., Computational Fluid Dynamics Modeling of Polymer Electrolyte Membrane Fuel Cells, J. Power Sources, 147, 95-106, (2005)

165. Inoue, G., Matsukuma, Y., Effect of Gas Channel Depth on Current Density Distribution of Polymer Electrolyte Fuel Cell by Numerical Analysis Including Gas Flow Through Gas Diffusion Layer, J. Power Sources, 157, 136-152, (2006)

166. Wu, C. W., Zhou, P., Ma, G. J., Influence of Clamping Force on the Performance of PEMFCs, J. Power Sources, 163, 874-881, (2007)

167. Wang, X., Shi, Z., Guessous, L., Effect of Compression on the Water Management of a Proton Exchange Membrane Fuel Cell With Different Gas Diffusion Layers, J. Fuel Cell Sci. Tech., 7/021012-1, (2010)

168. Li, X., Zhou, Y., Jiao, K., Du, Q., Yin, Y., Gas Diffusion Layer Deformation and Its Effect on the Transport Characteristics and Performance of Proton Exchange Membrane Fuel Cell, Int. J. Hydrogen Energy, 38, 12891.12903, (2013)

169. Al-Baghdadi, M. A. R. S., Al-Janabi, H. A. K., S., Effect of Operating Parameters on the Hygro-Thermal Stresses in Proton Exchange Membranes of Fuel Cells, Int. J. Hydrogen Energy, 32, 4510-4522, (2007)

170. Sainan, K. I., Atan, R., Husain, H., Mohamed, W. A. N. W., Computational Model Analysis on a Bipolar Plate Flow Field Design of a PEM Fuel Cell, Proceedings of 5th International Power Engineering and Optimization Conference (PEOCO2011), (2011)

171. Kim, S. H., Chun, J. H., Park, K. T., Jo, D. H., Kim, S. G., Numerical Modeling and Experimental Study of the Influence of GDL Properties on Performance in a PEMFC, Int. J. Hydrogen Energy, 36, 1837-1845, (2011)

172. Sui, P. C., Djilali, N., Fuel Cells-Proton-Exchange Membrane Fuel Cells Modeling, Encyclopedia of Electrochemical Power Sources, 868-878, (2009)

173. Son, Y., Lee, S., Jeong, H., Ahn, B., Lim, T., Parametric Study of The Channel Design at The Bipolar Plate in PEMFC Performances, J. Hydrogen Energy, 33, 5691-5696

174. Wu, C. W., Zhou, P., Numerical Study on the Compression Effect of Gas Diffusion Layer on PEMFC Performance, J. Power Sources, 170,93-100, (2007)

175. Chi, P. H., Chan, S. H., Weng, F. B., Su, A., Sui, P. C., Djilali, N., On the Effects of Non-uniform Property Distribution Due to Compression in the Gas Diffusionn Layer of a PEMFC, Int. J. Hydrogen Energy, 35, 2936-2948, (2010)

176. Ju, H., Chippar, P., O, Kyeongmin, Kang, K., A Numerical Investigation of the Effects of GDL Compression and Intrusion in Polymer Electrolyte Fuel Cells (PEFCs), Int. J. Hydrogen Energy, 37, 6326-6338, (2012)

177. Hottinen, T., Himanen, O., Karvonen, S., Nitta, I., Inhomogeneous Compression of PEMFC Gas Diffusion Layer: Part II. Modeling the Effect, J. Power Sources, 171, 113-121, (2007)

178. Al-Baghdadi, M. A. R. S., A CFD Study of Hygro-thermal Stresses Distribution in PEM Fuel Cell During Regular Cel Operation, Renew. Energ., 34, 674-682, (2009)

179. Serincan, M. F., Pasaogullari, U., Mechanical Behavior of the Membrane During the Polymer Electrolyte Fuel Cell Operation, J. Power Source, 196, 1303-1313, (2011)

180. Serincan, M. F., Pasaogullari, U., Effect of Gas Diffusion Layer anisotropy on Mechanical Stresses in a Polymer Electrolyte Membrane, J. Power Sources, 196, 1314-1320, (2011)

181. Taymaz, I., Benli, M., Numerical Study of Assembly Pressure Effect on the Performance of Proton Exchange Membrane Fuel Cell, Energy, 35, 2134-2140, (2010)

182. Hu, S. J., Zhou, Y., Lin, G., Shih, A. J., Assembly Pressure and Membrane Swelling in PEM Fuel Cells, J. Power Sources, 192, 544-551, (2009)

183. Su, Z. Y., Liu, C. T., Chang, H. P., Li, C. H., Huang, K. J., Sui, P. C., A Numerical Investigation of the Effects of Compression Force on PEM Fuel Cell Performance, J. Power Sources, 183, 182-192, (2008)

184. Wang, X., Shi, Z., A Numerical Study of Flow Crossover Between Adjacent Flow Channels in a Proton Exchange Membrane Fuel Cell with Serpentine Flow Field, J. Power Sources, 185, 985-992, (2008)

185. Lum, K. W., McGuirk, J. J., Three-dimensional Model of a Complete Polymer Electrolyte Membrane Fuel Cell-Model Formulation, Validation and Parametric Studies, J. Power Sources, 143, 103-124, (2005)

186. Reddy, R. G., Kumar, A., Effect of Channel Dimensions and Shape in the Flow-field Distributor on the Performance of Polymer Electrolyte Membrane Fuel Cells, J. Power Sources, 113, 11-18, (2003)

187. Djilali, N., Siversten, B. R., CFD-Based Modelling of Proton Exchange Membrane Fuel Cells, J. Power Sources, 141, 65-78, (2005)

188. Fan, J., Hu, G., Chen, S., Liu, Y., Cen, K., Three-dimensional Numerical Analysis of Proton Exchange Membrane Fuel Cells (PEMFCs) with Conventional and Interdigitated Flow Fields, J. Power Sources, 136, 1-9, (2004)

189. Brett, D. J. L., Daniels, F. A., Attingre, C., Kucemak, A. R., Current Collector Design for Closed-plenum Polymer Electrolyte Membrane Fuel Cells, J. Power Sources, 249, 247-262, (2014)

190. Brett, D. J. L., Daniels, F. A., Kucernak, A. R., Attingre, C., Using Computational Multiphysics to Optimise Channel Design for a Novel PEM Fuel Cell Stack, Proceedings of European COMSOL conference, Stuttgart, Germany, (2011)

191. Wang, X., Shi, Z., Zhang, Z., Comparison of Two-dimensional PEM Fuel Cell Modeling Using Comsol Multiphysics, Proceedings of COMSOL conference, Boston, USA, (2006)

192. Wang, Y., Um, S., Three-dimensional Analysis of Transport and Electrochemical Reactions in Polymer Electrolyte Fuel Cells, J. Power Sources, 125, 40-51, (2004)

193. Shan, Y., Choe, S. Y., Choi, S. H., Unsteady 2D PEM Fuel Cell Modeling for a Stack Emphasizing Thermal Effects, J. Power Sources, 165, 196-209, (2007)

194. Ju, H., Chippar, P., Oh, K., Numerical Study of Thermal Stress in High-Temperature Proton Exchange Membrane Fuel Cell (HT-PEMFC). Int. J. Hydrogen Energy, 39, 2785-2794, (2014)

195. Kvesić, Mirko, Modellierung und Simulation von Hochtemperatur-Polymerelektrolyt-Brennstoffzellen, RWTH Aachen University, Dissertation, (2012)

196. Siegel, C., Bandlamudi, G., Heinzel, A., Systematic Characterization of a PBI/H3PO4 sol-gel Membrane-Modeling and Simulation, J. Power Sources, 196, 2735-2749, (2011)

197. Comsol Batteries & Fuel Cell Module Model Library Manual, COMSOL Multiphysics® (4.3.b), (2013)

198. Broka, K., Ekdunge, P., J. Appl. Electrochem., 27, 281, (1997)

199. Dannenberg, K., Ekdunge, P., Lindenbergh, G., J. Appl. Electrochem., 30, 1377, (2000)

200. Nitta, I., Hottinen, T., Himanen, O., Mikkola, M., Inhomogeneous compression of PEMFC gas Diffusion Layer Part I. Experimental, J. Power Sources, 171, 26-36, (2007)

201. Brett, D. J. L., Mason, T. J.,Millichamp, J., Neville, T. P., El-Kharouri, A., Pollet, B. G., Effect of Clamping Pressure on Ohmic Resistance and Compression of Gas Diffusion Layers for Polymer Electrolyte Fuel Cells, J. Power Sources, 219, 52-59, (2012)

202. Park, J. W., Santamaria, A. D., Cooper, N. J., Becton, M. K., Effect of Channel Length on Interdigitated Flow-field PEMFC Performance: A Computational and Experimental Study, Int. J. Hydrogen Energy, 38, 16253-16263, (2013)

203. Squadrito, G., Barbera, O., Giacoppo, G., Urbani, F., Passalacqua, E., Polymer Electrolyte Fuel Cell Stack Research and Development, Int. J. Hydrogen Energy, 33, 1941-1946, (2008)

7 List of Figures

8   List of Tables

Lebenslauf

Persönliche Daten

Ahmet Evren Firat
geb. am 03.08.1982
In Erzincan, Türkei

Schulausbildung

1987-1992 Ege Sanayi Grundschule, Istanbul; Türkei
1992-1995 Yakacik Gymnasium, Istanbul, Türkei
1995-1999 Kadir Has Gymnasium, Istanbul, Türkei

Hochschulausbildung

10/1999-10/2000	Englisch für Ingenieure, Marmara Universität	Istanbul, Türkei
10/2000-10/2004	Maschinenbau (B.Sc.), Marmara Universität	Istanbul, Türkei
03/2005-04/2008	Maschinenbau (M.Sc.), Vertiefungsrichtung Mechatronik, Universität Duisburg-Essen	Duisburg

Berufliche Tätigkeiten

| 04/2008 | Wissenschaftlicher Mitarbeiter beim ZBT (Zentrum für BrennstoffzellenTechnik) | Duisburg |

Duisburg, im November 2015   Ahmet Evren Firat